HORTICULTURE: PRINCIPLES AND PRACTICES

Edited by
Hazelle S. Cabugao

ArclerPress

Horticulture: Principles and Practices

Edited by Hazelle S. Cabugao

Edition 2018

ArclerPress

Published by
Arcler Press LLC
708 3rd Avenue, 6th Floor
New York
NY 10017
United States of America
www.arclerpress.com

ISBN: 978-1-68094-551-5 (Hardback)

Printed in United States of America

Library of Congress Control Number: 2017933121

About the Editor

Hazelle S. Cabugao

Hazelle earned her degree in Agriculture from the University of the Philippines Los Baños in 2014. Her studies focus on ornamental horticulture particularly the use of plants and trees in designing landscapes. A horticulturist by profession, she supervises landscape implementation and maintenance and is also passionate about conserving the indigenous flora

Table of Contents

2 FLORICULTURE

3 THE ART OF ARBORICULTURE

6 PRACTICES OF OLERICULTURE

7 VITICULTURE

8 TROPICAL GARDENING

List of Abbreviations

ASLA	American Society of Landscape Architects
AVF	Association for Vertical Farming
AILA	Australian Institute of Landscape Architects
BOCLASA	Board of Control for Landscape Architects
CaMV	Cauliflower Mosaic Virus
CPD	Continuing Professional Development
CEA	controlled-Environment Agriculture
CLARB	Council of Landscape Architectural Registration Boards
CBN	Cubic Boron Nitride
CMS	Cytoplasmic Male Sterility
EFLA	European Federation for Landscape Architecture
GIS	Geographic Information Systems
HT-HP	High Temperature-High Pressure
HD	horticulture for Development
H.E.D	Human Ecology Design Team
ILASA	Institute for Landscape Architecture in South Africa
IOH	Institute of Horticulture
IFLA	International Federation of Landscape Architects
IVDMD	In Vitro Dry Matter Digestibility
ILI	Irish Landscape Institute
LI	Landscape Institute
NJHA	National Junior Horticultural Association
NZILA	New Zealand Institute of Landscape Architects
PPB	Participatory Plant Breeding

PDC	Polycrystalline Diamond Composite or Compacts
WBFF	World Bonsai Friendship Federation

Preface

The practice of horticulture can be retraced. It dates back to the times of Cyrus, the Great of ancient Persia, and has been going on ever since, with present-day horticulturists such as Freeman S. Howlett and Luther Burbank. The practice of horticulture can be retraced for many thousands of years. Horticulture is the science and art of producing, improving, marketing, and using fruits, vegetables, flowers, and ornamental plants. It differs from botany and other plant sciences in that horticulture incorporates both science and aesthetics. Production and consumption of high quality fruits and vegetables allows us to maintain a healthy, balanced daily diet. Flowers and ornamental plants enrich our homes and communities, and contribute to our sense of well-being. Horticulture impacts our lives on a daily basis by providing nutritious fruits and vegetables, offering visual enjoyment, and promoting recreational activities. Horticulture forms an integral and important component in the economy of a nation. Horticultural crops constitute a significant segment of the total agricultural production of a country. The importance of horticulture can be substantiated by its benefits like high export value, high per unit area yield, high returns per unit area, best utilization of wasteland, provision of raw materials for industries, whole engagement by a grower/laborer, production of more food energy per unit area than that of field crops, better use of undulating lands, and stabilization of women's empowerment by providing employment opportunities through processing, floriculture, seed production, mushroom cultivation, nursery preparation, etc. In addition, fruits and vegetables constitute the important energy-giving material to the human body. It also improves the economic condition of many farmers, and it has become a means of improving livelihood for many unprivileged classes too. Flower harvesting, nursery maintenance, hybrid seed production and tissue culture, propagation of fruits and flowers, and food processing are highly remunerative employment options for women in rural areas. It has also been common throughout history for people to raise fruits and vegetables in small gardens to supplement their diets. A "green revolution" took place in the 1960s and 1970s, and the interest in horticulture grew. Today, horticulture is popular and important in all societies, both rich and poor. The popularity of horticulture is due to a number of factors. The horticulture industry provides economic value to local communities and to the country. A healthy industry creates job opportunities. Horticultural practices provide high-quality food for people. It also offers offers an increase in aesthetic pleasure that supports psychological well-being. This book offers the comprehensive accessible information available on indoor and outdoor gardening.

Content Coverage

Chapter One is an overview of horticulture, such as etymology, history, and scope.

Chapter Two aims to provide a comprehensive overview of floriculture cultivation of flowering and ornamental plants for gardens and for floristry.

Chapter Three discusses the practices of arboriculture including selecting trees for specific purposes, planting them, and caring for them.

Chapter Four is an overview of landscape architecture which is related to both artistic and scientific aspects of landscaping and plant selection.

Chapters Five and *Six* present about garden tools and practices of olericulture. Olericulture is the science of vegetable growing, dealing with the culture of non-woody plants for food.

Chapter Seven is about viticulture; the science, production, and study of grapes.

Chapter Eight discusses about how to cultivate the plants in the tropics.

CHAPTER 1

Horticulture: An Overview

Chapter Contents

Horticulture is the art and science of cultivating and managing crops, gardens, plants, flowers and trees. Not to be confused with botany, the scientific study of plantlife, horticulture encompasses a broad spectrum of areas, from crop production to flower arrangement.

INTRODUCTION

Horticulture is the branch of agriculture that deals with the art, science, technology, and business of growing plants. It includes the cultivation of medicinal plants, fruits, vegetables, nuts, seeds, herbs, sprouts, mushrooms, algae, flowers, seaweeds and non-food crops such as grass and ornamental trees and plants. It also includes plant conservation, landscape restoration, landscape and garden design, construction, and maintenance, and arboriculture. Inside agriculture, horticulture contrasts with extensive field farming as well as animal husbandry. Horticulturists apply their knowledge, skills, and technologies used to grow intensively produced plants for human food and non-food uses and for personal or social needs. Their work involves plant propagation and cultivation with the aim of improving plant growth, yields, quality, nutritional value, and resistance to insects, diseases, and environmental stresses. They work as gardeners, growers, therapists, designers, and technical advisors in the food and non-food sectors of horticulture. Horticulture even refers to the growing of plants in a field or garden.

A fruit and vegetable market in Milan

A vertical garden

Horticultural and health-related produce

Etymology

The word horticulture is modeled after agriculture, and comes from the Greek χόρτος, which in Latin became hortus "garden" and cultūra "cultivation", from cultus, the perfect passive participle of the verb colō "I cultivate". Hortus is cognate with the native English word yard (in the meaning of land associated with a building) and also the borrowed word garden.

Scope

Horticulture involves nine areas of study, which can be grouped into two broad sections, ornamentals and edibles:

- Arboriculture is the study of and the selection, establishment, care, and removal of individual trees, shrubs, vines, and other perennial woody plants.
- Turf management includes all aspects of the production and maintenance of turf grass for sports, leisure use or amenity use.
- Floriculture includes the production and marketing of floral crops.
- Landscape horticulture includes the production, marketing and maintenance of landscape plants.
- Olericulture includes the production and marketing of vegetables.
- Pomology includes the production and marketing of pome fruits.
- Viticulture includes the production and marketing of grapes.
- Oenology includes all aspects of wine and winemaking.

- Postharvest physiology involves maintaining the quality of and preventing the spoilage of plants and animals.

Anthropology

Horticulture has a very long history. The study and science of horticulture dates all the way back to the times of Cyrus the Great of ancient Persia, and has been going on ever since, with present-day horticulturists such as Freeman S. Howlett and Luther Burbank. The practice of horticulture can be retraced for many thousands of years. The cultivation of taro and yam in Papua New Guinea dates back to at least 6950–6440 cal BP.The origins of horticulture lie in the transition of human communities from nomadic hunter-gatherers to sedentary or semi-sedentary horticultural communities, cultivating a variety of crops on a small scale around their dwellings or in specialized plots visited occasionally during migrations from one area to the next (such as the "milpa" or maize field of Mesoamerican cultures). In the Pre-Columbian Amazon Rainforest, natives are believed to have used biochar to enhance soil productivity by smoldering plant waste. European settlers called it Terra Preta de Indio. In forest areas such horticulture is often carried out in swiddens ("slash and burn" areas). A characteristic of horticultural communities is that useful trees are often to be found planted around communities or specially retained from the natural ecosystem.

Horticulture primarily differs from agriculture in two ways. First, it generally encompasses a smaller scale of cultivation, using small plots of mixed crops rather than large fields of single crops. Secondly, horticultural cultivations generally include a wide variety of crops, even including fruit trees with ground crops. Agricultural cultivations however as a rule focus on one primary crop. In pre-contact North America the semi-sedentary horticultural communities of the Eastern Woodlands (growing maize, squash and sunflower) contrasted markedly with the mobile hunter-gatherer communities of the Plains people. In Central America, Maya horticulture involved augmentation of the forest with useful trees such as papaya, avocado, cacao, ceiba and sapodilla. In the cornfields, multiple crops were grown such as beans (using cornstalks as supports), squash, pumpkins and chilli peppers. In some cultures these were tended mainly or exclusively by women.

Organizations

The oldest Horticultural society in the world, founded in 1768, is the Ancient Society of York Florists. They still have four shows a year in York, UK.They have a large archive of records, including the original members book dating back to 1768.

Desh Bhagat School of Horticulture Sciences at Desh Bhagat University, Punjab is a prestigious institution in India to encourage the research and promote horticulture techniques throughout the country. The Royal Horticultural Society is a UK charity which oversees several major shows and gardens.

The professional body representing horticulturists in Great Britain and Ireland is the Institute of Horticulture (IOH). Also, the IOH has an international branch for members outside of these islands. The International Society for Horticultural Science promotes and encourages research and education in all branches of horticultural science.

The American Society of Horticultural Science promotes and encourages research and education in all branches of horticultural science in the Americas.

The Australian Society of Horticultural Science was established in 1990 as a professional society for the promotion and enhancement of Australian horticultural science and industry.

The National Junior Horticultural Association (NJHA) was established in 1934 and was the first organization in the world dedicated solely to youth and horticulture. NJHA programs are designed to help young people obtain a basic understanding of, and develop skills in, the ever expanding art and science of horticulture.

The New Zealand Horticulture Institute.

The Global Horticulture Initiative (GlobalHort) fosters more efficient and effective partnerships and collective action among different stakeholders in horticulture. The organization has a special focus on horticulture for development (H4D), i.e. using horticulture to reduce poverty and improve nutrition worldwide. To be efficient, GlobalHort is organized in a consortium of national and international organizations to collaborate in research, training, and technology-generating activities designed to meet mutually-agreed-upon objectives. GlobalHort is a non-for-profit organization registered in Belgium.

GARDENING

Part of a parterre in an English garden

Gardening is the practice of growing and cultivating plants as part of horticulture. In gardens, ornamental plants are often grown for their flowers, foliage, or overall appearance; useful plants, such as root vegetables, leaf vegetables, fruits, and herbs, are grown for consumption, for use as dyes, or for medicinal and cosmetic use. Gardening is considered to be a relaxing activity for many people.

Gardening ranges in scale from fruit orchards, to long boulevard plantings with one or more different types of shrubs, trees, and herbaceous plants, to residential yards including lawns and foundation plantings, to plants in large or small containers grown inside or outside. Gardening may be very specialized, with only one type of plant grown, or involve a large number of different plants in mixed plantings. It involves an active participation in the growing of plants, and tends to be labor-intensive, which differentiates it from farming or forestry.

Ancient times

Forest gardening, a forest-based food production system, is the world's oldest form of gardening. Forest gardens originated in prehistoric times along jungle-clad river banks and in the wet foothills of monsoon regions. In the gradual process of families improving their immediate environment, useful tree and vine species were identified, protected and improved while undesirable species were eliminated. Eventually foreign species were also selected and incorporated into the gardens.

After the emergence of the first civilizations, wealthy individuals began to create gardens for aesthetic purposes. Egyptian tomb paintings from around 1500 BC provide some of the earliest physical evidence of ornamental horticulture and landscape design; they depict lotus ponds surrounded by symmetrical rows of acacias and palms. A notable example of ancient ornamental gardens were the Hanging Gardens of Babylon—one of the Seven Wonders of the Ancient World —while ancient Rome had dozens of gardens.

Wealthy ancient Egyptians used gardens for providing shade. Egyptians associated trees and gardens with gods as they believed that their deities were pleased by gardens. Gardens in ancient Egypt were often surrounded by walls with trees planted in rows. Among the most popular species planted were date palms, sycamores, fir trees, nut trees, and willows. These gardens were a sign of higher socioeconomic status. In addition, wealthy ancient Egyptians grew vineyards, as wine was a sign of the higher social classes. Roses, poppies, daisies and irises could all also be found in the gardens of the Egyptians.

Assyria was also renowned for its beautiful gardens. These tended to be wide and large, some of them used for hunting game—rather like a game reserve today—and others as leisure gardens. Cypresses and palms were some of the most frequently planted types of trees.

Ancient Roman gardens were laid out with hedges and vines and contained a wide variety of flowers—acanthus, cornflowers, crocus, cyclamen, hyacinth, iris, ivy, lavender, lilies, myrtle, narcissus, poppy, rosemary and violets—as well as statues and sculptures. Flower beds were popular in the courtyards of rich Romans.

A gardener at work, 1607

The Middle Ages

The Middle Age represented a period of decline in gardens for aesthetic purposes, with regard to gardening. After the fall of Rome, gardening was done for the purpose of growing medicinal herbs and/or decorating church altars. Monasteries carried on a tradition of garden design and intense horticultural techniques during the medieval period in Europe. Generally, monastic garden types consisted of kitchen gardens, infirmary gardens, cemetery orchards, cloister garths and vineyards. Individual monasteries might also have had a “green court”, a plot of grass and trees where horses could graze, as well as a cellarer’s garden or private gardens for obedientiaries, monks who held specific posts within the monastery.

Islamic gardens were built after the model of Persian gardens and they were usually enclosed by walls and divided in 4 by watercourses. Commonly, the center of the garden would have a pool or pavilion. Specific to the Islamic gardens are the mosaics and glazed tiles used to decorate the rills and fountains that were built in these gardens.

By the late 13th century, rich Europeans began to grow gardens for leisure and for medicinal herbs and vegetables. They surrounded the gardens by walls to protect them from animals and to provide seclusion. During the next two centuries, Europeans started planting lawns and raising flowerbeds and trellises of roses. Fruit trees were common in these gardens and also in some, there were turf seats. At the same time, the gardens in the monasteries were a place to grow flowers and medicinal herbs but they were also a space where the monks could enjoy nature and relax.

The gardens in the 16th and 17th century were symmetric, proportioned and balanced with a more classical appearance. Most of these gardens were built around a central axis and they were divided into different parts by hedges. Commonly, gardens had flowerbeds laid out in squares and separated by gravel paths.

Gardens in Renaissance were adorned with sculptures, topiary and fountains. In the 17th century, knot gardens became popular along with the hedge mazes. By this time, Europeans started planting new flowers such as tulips, marigolds and sunflowers.

Cottage Gardens

Cottage gardens, which emerged in Elizabethan times, appear to have originated as a local source for herbs and fruits. One theory is that they arose out of the Black Death of the 1340s, when the death of so many laborers made land available for small cottages with personal gardens.According to the late 19th-century legend of origin, these gardens were originally created by the workers that lived in the cottages of the villages, to provide them with food and herbs, with flowers planted among them for decoration. Farm workers were provided with cottages that had architectural quality set in a small garden—about an acre—where they could grow food and keep pigs and chickens.

Authentic gardens of the yeoman cottager would have included a beehive and livestock, and frequently a pig and sty, along with a well. The peasant cottager of medieval times was more interested in meat than flowers, with herbs grown for medicinal use rather than for their beauty. By Elizabethan times there was more prosperity, and thus more room to grow flowers. Even the early cottage garden flowers typically had their practical use—violets were spread on the floor (for their pleasant scent and keeping out vermin); calendulas and primroses were both attractive and used in cooking. Others, such as sweet william and hollyhocks, were grown entirely for their beauty.

18th Century

In the 18th century, gardens were laid out more naturally, without any walls. This style of smooth undulating grass, which would run straight to the house, clumps, belts and scattering of trees and his serpentine lakes formed by invisibly damming small rivers, were a new style within the English landscape, a "gardenless" form of landscape gardening, which swept away almost all the remnants of previous formally patterned styles. The English garden usually included a lake, lawns set against groves of trees, and often contained shrubberies, grottoes, pavilions, bridges and follies such as mock temples, Gothic ruins, bridges, and other picturesque architecture, designed to recreate an idyllic pastoral landscape. This new style emerged in England in the early 18th century, and spread across Europe, replacing the more formal, symmetrical Garden à la française of the 17th century as the principal gardening style of Europe. The English garden presented an idealized view of nature. They were often inspired by paintings of landscapes by Claude Lorraine and Nicolas Poussin, and some were Influenced by the classic Chinese gardens of the

East, which had recently been described by European travelers. The work of Lancelot ‹Capability` Brown was particularly influential. Also, in 1804 the Horticultural Society was formed. Gardens of the 19th century contained plants such as the monkey puzzle or Chile pine. This is also the time when the so-called "gardenesque" style of gardens evolved. These gardens displayed a wide variety of flowers in a rather small space. Rock gardens increased in popularity in the 19th century.

Types

Conservatory of Flowers in Golden Gate Park, San Francisco

Hanging baskets in Thornbury, South Gloucestershire

Residential gardening takes place near the home, in a space referred to as the **garden**. Although a garden typically is located on the land near a residence, it may also be located on a roof, in an atrium, on a balcony, in a windowbox, or on a patio or vivarium.

Gardening also takes place in non-residential green areas such as parks, public or semi-public gardens (botanical gardens or zoological gardens), amusement parks, along transportation corridors, and around tourist attractions and garden hotels. In these situations, a staff of gardeners or groundskeepers maintains the gardens.

- Indoor gardening is concerned with the growing of houseplants within a residence or building, in a conservatory, or in a greenhouse. Indoor gardens are sometimes incorporated as part of air conditioning or heating systems. Indoor gardening extends the growing season in the fall and spring and can be used for winter gardening.
- Native plant gardening is concerned with the use of native plants with or without the intent of creating wildlife habitat. The goal is to create a garden in harmony with, and adapted to a given area. This type of gardening typically reduces water usage, maintenance, and fertilization costs, while increasing native faunal interest.
- Water gardening is concerned with growing plants adapted to pools and ponds. Bog gardens are also considered a type of water garden. These all require special conditions and considerations. A simple water garden may consist solely of a tub containing the water and plant(s). In aquascaping, a garden is created within an aquarium tank.
- Container gardening is concerned with growing plants in any type of container either indoors or outdoors. Common containers are pots, hanging baskets, and planters. Container gardening is usually used in atriums and on balconies, patios, and roof tops.
- Hügelkultur is concerned with growing plants on piles of rotting wood, as a form of raised bed gardening and composting in situ. An English loanword from German, it means "mound garden." Toby Hemenway, noted Permaculture author and teacher, considers wood buried in trenches to also be a form of hugelkultur referred to as a dead wood swale. Hugelkultur is practiced by Sepp Holzer as a method of forest gardening and agroforestry, and by Geoff Lawton as a method of dryland farming and desert greening. When used as a method of disposing large volumes of waste wood and woody debris, hugelkultur accomplishes carbon sequestration. It is also a form of xeriscaping.
- Community gardening is a social activity in which an area of land is gardened by a group of people, providing access to fresh produce and plants as well as access to satisfying labor, neighborhood improvement, sense of community and connection to the environment.Community gardens are typically owned in trust by local governments or nonprofits.
- Garden sharing partners landowners with gardeners in need of land. These shared gardens, typically front or back yards, are usually used to produce food that is divided between the two parties.

- Organic gardening uses natural, sustainable methods, fertilizers and pesticides to grow non-genetically modified crops.

Garden Features and Accessories

There is a wide range of features and accessories available in the market for both the professional gardener and the amateur to exercise their creativity. These are used to add decoration or functionality, and may be made from a wide range of materials such as copper, stone, wood, bamboo, stainless steel, clay, stained glass, concrete, or iron. Examples include trellis, arbors, statues, benches, water fountains, urns, bird baths and feeders, and garden lighting such as candle lanterns and oil lamps. The use of these items can be part of the expression of a gardener's gardening personality.

Gardening Departments and Centers

Gardening departments and centers mainly sell plants, sundries, and garden accessories, but nowadays, many stock outdoor leisure products as diverse as spas, furniture, and barbecues. Many garden centers now include food halls and sections for clothing, gifts, pets, and power tools. There are also a number of online garden centers that deliver directly to customers' doors.

Comparison with Farming

Hand gardening tools

Gardening for beauty is likely nearly as old as farming for food, however for most of history for the majority of people there was no real distinction since the need for food and other useful product trumped other concerns. Small-scale, subsistence agriculture (called hoe-farming) is largely indistinguishable from gardening. A patch of potatoes grown by a Peruvian peasant or an Irish smallholder for personal use could be described as either a garden or a farm. Gardening for average people evolved as a separate discipline, more concerned with aesthetics and recreation, under the influence of the pleasure gardens of the wealthy. Meanwhile, farming has evolved (in developed countries) in the direction of commercialization, economics of scale, and monocropping.

In respect to its food producing purpose, gardening is distinguished from farming chiefly by scale and intent. Farming occurs on a larger scale, and with the production of salable goods as a major motivation. Gardening is done on a smaller scale, primarily for pleasure and to produce goods for the gardener's own family or community. There is some overlap between the terms, particularly in that some moderate-sized vegetable growing concerns, often called market gardening, can fit in either category.

Planting in a garden

Gardening can be a hobby or an income supplement, but farming is generally understood as a full-time or commercial activity, usually involving more land and quite different practices. One distinction is that gardening is labor-intensive and employs very little infrastructural capital, sometimes no more than a few tools, e.g. a spade, hoe, basket and watering can. By contrast, larger-scale farming often involves irrigation systems, chemical fertilizers and harvesters or at least ladders, e.g. to reach up into fruit trees. However, this distinction is becoming blurred with the increasing use of power tools in even small gardens.

In part because of labor intensity and aesthetic motivations, gardening is very often much more productive per unit of land than farming. In the Soviet Union, half the food supply came from small peasants' garden plots on the huge government-run collective farms, although they were tiny patches of land. Some argue this as evidence of the superiority of capitalism, since the peasants were generally able to sell their produce.

Others consider it to be evidence of a tragedy of the commons, since the large collective plots were often neglected, or fertilizers or water redirected to the private gardens.

The term precision agriculture is sometimes used to describe gardening using intermediate technology (more than tools, less than harvesters), especially of organic varieties. Gardening is effectively scaled up to feed entire villages of over 100 people from specialized plots. A variant is the community garden which offers plots to urban dwellers.

Gardens as Art

Garden at the Schultenhof in Mettingen, North Rhine-Westphalia, Germany

Garden design is considered to be an art in most cultures, distinguished from gardening, which generally means garden maintenance. Garden design can include different themes such as perennial, butterfly, wildlife, Japanese, water, tropical, or shade gardens. In Japan, Samurai and Zen monks were often required to build decorative gardens or practice related skills like flower arrangement known as ikebana. In 18th-century Europe, country estates were refashioned by landscape gardeners into formal gardens or landscaped park lands, such as at Versailles, France, or Stowe, England. Today, landscape architects and garden designers continue to produce artistically creative designs for private garden spaces. In the USA, professional landscape designers are certified by the Association of Professional Landscape Designers.

Social Aspects

People can express their political or social views in gardens, intentionally or not. The lawn vs. garden issue is played out in urban planning as the debate over the "land ethic" that is to determine urban land use and whether hyper hygienist bylaws (e.g. weed control)

should apply, or whether land should generally be allowed to exist in its natural wild state. In a famous Canadian Charter of Rights case, "Sandra Bell vs. City of Toronto", 1997, the right to cultivate all native species, even most varieties deemed noxious or allergenic, was upheld as part of the right of free expression.

Community gardening comprises a wide variety of approaches to sharing land and gardens.

People often surround their house and garden with a hedge. Common hedge plants are privet, hawthorn, beech, yew, leyland cypress, hemlock, arborvitae, barberry, box, holly, oleander, forsythia and lavender. The idea of open gardens without hedges may be distasteful to those who enjoy privacy. The Slow Food movement has sought in some countries to add an edible school yard and garden classrooms to schools, e.g. in Fergus, Ontario, where these were added to a public school to augment the kitchen classroom. Garden sharing, where urban landowners allow gardeners to grow on their property in exchange for a share of the harvest, is associated with the desire to control the quality of one's food, and reconnect with soil and community.

In US and British usage, the production of ornamental plantings around buildings is called landscaping, landscape maintenance or grounds keeping, while international usage uses the term gardening for these same activities.

Also gaining popularity is the concept of "Green Gardening" which involves growing plants using organic fertilizers and pesticides so that the gardening process - or the flowers and fruits produced thereby - doesn`t adversely affect the environment or people`s health in any manner.

Garden Pests

Garden pests are generally plants, fungi, or animals (frequently insects) that engages in activity that the gardener considers undesirable. It may crowd out desirable plants, disturb soil, stunt the growth of young seedlings, steal or damage fruit, or otherwise kill plants, hamper their growth, damage their appearance, or reduce the quality of the edible or ornamental portions of the plant. Aphids, spider mites, slugs, snails, ants, birds, weeds and even cats are commonly considered to be garden pests.

Because gardeners may have different goals, organisms considered "garden pests" vary from gardener to gardener. Tropaeolum speciosum, for example, may be considered a desirable and ornamental garden plant, or it may be considered a pest if it seeds and starts to grow where it is not wanted. As another example, in lawns, moss can become dominant and be impossible to eradicate. In some lawns, lichens, especially very damp lawn lichens such as Peltigera lactucfolia and P. membranacea, can become difficult to control and be considered pests.

Garden Pest Control

There are many ways by which unwanted pests are removed from a garden. The techniques vary depending on the pest, the gardener's goals, and the gardener's philosophy. For example, snails may be dealt with through the use of a chemical pesticide, an organic pesticide, hand-picking, barriers, or simply growing snail-resistant plants. Pest control is often done through the use of pesticides, which may be either organic or artificially synthesized. Pesticides may affect the ecology of a garden due to their effects on the populations of both target and non-target species. For example, unintended exposure to some neonicotinoid pesticides has been proposed as a factor in the recent decline in honey bee populations. A mole vibrator can deter mole activity in a garden.

Other means of control include the removal of infected plants, using fertilizers and biostimulants to improve the health and vigour of plants so they better resist attack, practising crop rotation to prevent pest build-up, using companion planting, and practising good garden hygiene, such as disinfecting tools and clearing debris and weeds which may harbour pests.

VERTICAL FARMING

Vertical farming is the practice of producing food in vertically stacked layers, such as in a skyscraper, used warehouse, or shipping container. The modern ideas of vertical farming use indoor farming techniques and controlled-environment agriculture (CEA) technology, where all environmental factors can be controlled. These facilities utilize artificial control of light, environmental control (humidity, temperature, gases...) and fertigation. Some vertical farms use techniques similar to greenhouses, where natural sunlight can be augmented with artificial lighting and metal reflectors.

Lettuce grown in indoor vertical farming system.

Types

The term "vertical farming" was coined by Gilbert Ellis Bailey in 1915 in his book Vertical Farming. His use of the term differs from the current meaning—he wrote about farming with a special interest in soil origin, its nutrient content and the view of plant life as "vertical" life forms, specifically relating to their underground root structures. Modern usage of the term "vertical farming" usually refers to growing plants in layers, whether in a multistory skyscraper, used warehouse, or shipping container.

Mixed-Use Skyscrapers

Mixed-use skyscrapers were proposed and built by architect Ken Yeang. Yeang proposes that instead of hermetically sealed mass-produced agriculture, plant life should be cultivated within open air, mixed-use skyscrapers for climate control and consumption. This version of vertical farming is based upon personal or community use rather than the wholesale production and distribution plant life that aspires to feed an entire city. It thus requires less of an initial investment than Despommier's "vertical farm". However, neither Despommier nor Yeang are the conceptual originators, nor is Yeang the inventor of vertical farming in skyscrapers.

Despommier's Skyscrapers

Dickson Despommier shares his ideas about how "vertical farming" can help reduce hunger by changing the way we use land for agriculture. Photography by Kris Krüg

Ecologist Dickson Despommier argues that vertical farming is legitimate for environmental reasons. He claims that the cultivation of plant life within skyscrapers will require less embedded energy and produce less pollution than some methods of producing plant life on natural landscapes. He moreover claims that natural landscapes are too toxic for natural, agricultural production, despite the ecological and environmental costs of extracting materials to build skyscrapers for the simple purpose of agricultural production.

Vertical farming according to Despommier thus discounts the value of natural landscape in exchange for the idea of "skyscraper as spaceship." Plant life is mass-produced within hermetically sealed, artificial environments that have little to do with the outside world. In this sense, they could be built anywhere regardless of the context. Although climate control, lighting, and other costs of maintenance have been posited as potentially stifling to bringing this concept to fruition, advocates have countered that an important feature of future vertical farms will be the integration of renewable energy technology, be it solar panels, wind turbines, water capture systems, and probably some combination of the three. The vertical farm is designed to be sustainable, and to enable nearby inhabitants to work at the farm.

Despommier`s concept of the vertical farm emerged in 1999 at Columbia University. It promotes the mass cultivation of plant life for commercial purposes in skyscrapers.

Stackable Shipping Containers

Several companies have brought forth the concept of stacking recycled shipping containers in urban settings. Freight Farms produces a "leafy green machine" that is a complete farm-to-table system outfitted with vertical hydroponics, LED lighting and intuitive climate controls built within a 12 m × 2.4 m shipping container. Podponics has built a large scale vertical farm in Atlanta consisting of over 100 stacked "growpods". A similar farm is currently under construction in Oman.

Problems

Economics

Opponents question the potential profitability of vertical farming. A detailed cost analysis of start-up costs, operation costs and revenue has not been done. The extra cost of lighting, heating, and powering the vertical farm may negate any of the cost benefits received by the decrease in transportation expenses. The economic and environmental benefits of vertical farming rest partly on the concept of minimizing food miles, the distance that food travels from farm to consumer. However, a recent analysis suggests that transportation is only a minor contributor to the economic and environmental costs of supplying food to urban populations. The author of the report, University of Toronto professor Pierre Desrochers, concluded that "food miles are, at best, a marketing fad." Thus the facility would have to produce a considerable profit to justify remaining in the city. A simpler concept rather than trying to stack farms on multiple stories would be to just cultivate crops on the roofs of existing building. Rooftop farming is a growing urban trend, requires little construction (other than fortifying the roof to hold the weight of the growing medium), still takes advantage of sunlight and doesn't require investment in machinery, growing lights or irrigation.

Similarly, if the power needs of the vertical farm are met by fossil fuels, the environmental effect may be a net loss; even building low-carbon capacity to power the farms may not make as much sense as simply leaving the traditional farms in place, and burning less coal. The initial building costs will be easily over $100 million, for a 60 hectare vertical farm. Office occupancy costs can be very high in major cities, with cities such as Tokyo, Moscow, Mumbai, Dubai, Milan, Zurich, and Sao Paulo ranging from $1850 to $880 per square meter, respectively.

Energy Use

During the growing season, the sun shines on a vertical surface at an extreme angle such that much less light is available to crops than when they are planted on flat land. Therefore, supplemental light would be required in order to obtain economically viable yields. Bruce Bugbee, a crop physiologist at Utah State University, believes that the power demands of vertical farming will be too expensive and uncompetitive with traditional farms using only free natural light. The environmental writer George Monbiot calculated that the cost of providing enough supplementary light to grow the grain for a single loaf would be about $15. An article in the Economist argued that "even though crops growing in a glass skyscraper will get some natural sunlight during the day, it won't be enough" and "the cost of powering artificial lights will make indoor farming prohibitively expensive".

As "The Vertical Farm" proposes a controlled environment, heating and cooling costs will be at least as costly as any other tower. But there also remains the issue of complicated, if not more expensive, plumbing and elevator systems to distribute food and water throughout. Even throughout the northern continental United States, while heating with relatively cheap fossil fuels, the heating cost can be over $200,000 per hectare.

To address this problem, The Plant in Chicago is building an anaerobic digester into the building. This will allow the farm to operate off the energy grid. Moreover, the anaerobic digester will be recycling waste from nearby businesses that would otherwise go into landfills.

Pollution

Depending on the method of electricity generation used, regular greenhouse produce can create more greenhouse gases than field produce, largely due to higher energy use per kilogram of produce. With vertical farms requiring much greater energy per kilogram of produce, mainly through increased lighting, than regular greenhouses, the amount of pollution created will be much higher than that from field produce. The amount of pollution produced is dependent on how the energy used in the process is generated.

As plants acquire nearly all their carbon from the atmosphere, greenhouse growers commonly supplement CO_2 levels 3–4 times the rate normally found in the atmosphere. This increase in CO_2, which has been shown to increase photosynthesis rates by 50%, contributes to the higher yields expected in vertical farming. It is not uncommon to find greenhouses burning fossil fuels purely for this purpose, as other CO_2 sources, like from furnaces, contain pollutants such as sulphur dioxide and ethylene which significantly damage plants. This means a vertical farm will require a CO_2 source, most likely from combustion, even if the rest of the farm is powered by "green" energy. Also, through necessary ventilation, much CO_2 will be leaked into the city's atmosphere.

Greenhouse growers commonly exploit photoperiodism in plants to control whether the plants are in a vegetative or reproductive stage. As part of this control, growers will have the lights on past sunset and before sunrise or periodically throughout the night. Single story greenhouses are already a nuisance to neighbours because of light pollution, a 30-story vertical farm in a densely populated area will surely face problems because of its light pollution.

Hydroponics greenhouses regularly change the water, meaning there is a large quantity of water containing fertilizers and pesticides that must be disposed of. While solutions are currently being worked on, the most common method of simply spreading the mixture over a sufficient area of neighbouring farmland or wetlands would be more difficult for an urban vertical farm.

Advantages

Several potential advantages of vertical farming have been discussed by Despommier. Many of these benefits are obtained from scaling up hydroponic or aeroponic growing methods.

Preparation for the Future

It is estimated that by the year 2050, close to 80% of the world's population will live in urban areas and the total population of the world will increase by 3 billion people. A very large amount of land may be required depending on the change in yield per hectare. Scientists are concerned that this large amount of required farmland will not be available and that severe damage to the earth will be caused by the added farmland. According to Despommier, vertical farms, if designed properly, may eliminate the need to create additional farmland and help create a cleaner environment.

Increased Crop Production

Unlike traditional farming in non-tropical areas, indoor farming can produce crops year-round. All-season farming multiplies the productivity of the farmed surface by a factor of 4 to 6 depending on the crop. With some crops, such as strawberries, the factor may be as high as 30.

Furthermore, as the crops would be sold in the same infrastructures in which they are grown, they will not need to be transported between production and sale, resulting in less spoilage, infestation, and energy required than conventional farming encounters. Research has shown that 30% of harvested crops are wasted due to spoilage and infestation, though this number is much lower in developed nations.

Despommier suggests that, if dwarf versions of certain crops are used (e.g. dwarf wheat, which has been grown in space by NASA, is smaller in size but richer in nutrients), year-round crops, and "stacker" plant holders are accounted for, a 30-story building with a base of a building block (2 hectares (5 acres)) would yield a yearly crop analogous to that of 1,000 hectares (2,400 acres) of traditional farming.

Protection from Weather-Related Problems

Crops grown in traditional outdoor farming suffer from the often suboptimal, and sometimes extreme, nature of geological and meteorological events such as undesirable temperatures or rainfall amounts, monsoons, hailstorms, tornadoes, flooding, wildfires, and severe droughts. The protection of crops from weather is increasingly important as global climate change occurs. "Three recent floods (in 1993, 2007 and 2008) cost the United States billions of dollars in lost crops, with even more devastating losses in topsoil. Changes in rain patterns and temperature could diminish India's agricultural output by 30 percent by the end of the century."

Because vertical plant farming provides a controlled environment, the productivity of vertical farms would be mostly independent of weather and protected from extreme weather events. Although the controlled environment of vertical farming negates most of these factors, earthquakes and tornadoes still pose threats to the proposed infrastructure, although this again depends on the location of the vertical farms.

Conservation of Resources

Each unit of area in a vertical farm could allow up to 20 units of area of outdoor farmland to return to its natural state, and recover farmlands due to development from original flat farmlands.

Vertical farming would reduce the need for new farmland due to overpopulation, thus saving many natural resources, currently threatened by deforestation or pollution. Deforestation and desertification caused by agricultural encroachment on natural biomes would be avoided. Because vertical farming lets crops be grown closer to consumers, it would substantially reduce the amount of fossil fuels currently used to transport and refrigerate farm produce. Producing food indoors reduces or eliminates conventional plowing, planting, and harvesting by farm machinery, also powered by fossil fuels.

Halting Mass Extinction

Withdrawing human activity from large areas of the Earth's land surface may be necessary to slow and eventually halt the current anthropogenic mass extinction of land animals.

Traditional agriculture is highly disruptive to wild animal populations that live in and around farmland and some argue it becomes unethical when there is a viable alternative. One study showed that wood mouse populations dropped from 25 per hectare to 5 per hectare after harvest, estimating 10 animals killed per hectare each year with conventional farming. In comparison, vertical farming would cause very little harm to wildlife, and would allow disused farmland to return to its pre-agricultural state.

Impact on Human Health

Traditional farming is a hazardous occupation with particular risks that often take their toll on the health of human laborers. Such risks include: exposure to infectious diseases such as malaria and schistosomes, exposure to toxic chemicals commonly used as pesticides and fungicides, confrontations with dangerous wildlife such as venomous snakes, and the severe injuries that can occur when using large industrial farming equipment. Whereas the traditional farming environment inevitably contains these risks (particularly in the farming practice known as "slash and burn"), vertical farming – because the environment is strictly controlled and predictable – reduces some of these dangers. Currently, the American food system makes fast, unhealthy food cheap while fresh produce is less available and more expensive, encouraging poor eating habits. These poor eating habits lead to health problems such as obesity, heart disease, and diabetes. The increased availability and subsequent lower cost of fresh produce would encourage healthier eating.

Poverty/Destitution and Culture

Food security is one of the primary factors leading to absolute poverty. Being able to construct 'farm land' in secure areas as needed will help alleviate the pressures causing crises among neighbors fighting for resources (mainly water and space). It also allows continued growth of culturally significant food items without sacrificing sustainability or basic needs, which can be significant to the recovery of a society from poverty.

Urban Growth

Vertical farming, used in conjunction with other technologies and socioeconomic practices, could allow cities to expand while remaining largely self-sufficient food wise. This would allow for large urban centers that could grow without destroying considerably larger areas of forest to provide food for their people. Moreover, the industry of vertical farming will provide employment to these expanding urban centers. This may help displace the unemployment created by the dismantling of traditional farms, as more farm laborers move to cities in search of work.

Energy Sustainability

Vertical farms could exploit methane digesters to generate a small portion of its own electrical needs. Methane digesters could be built on site to transform the organic waste generated at the farm into biogas which is generally composed of 65% methane along with other gases. This biogas could then be burned to generate electricity for the greenhouse.

Technologies and Devices

Vertical farming relies on the use of various physical methods to become effective. Combining these technologies and devices in an integrated whole is necessary to make Vertical Farming a reality. Various methods are proposed and under research. The most common technologies suggested are:

- Greenhouse
- The Folkewall and other vertical growing architectures
- Flowerpot
- Aeroponics / Hydroponics / Aquaponics

- Composting
- Grow light
- Phytoremediation
- Skyscraper
- Controlled-environment agriculture
- Precision agriculture
- Agricultural robot

Plans

Despommier argues that the technology to construct vertical farms currently exists. He also states that the system can be profitable and effective, a claim evidenced by some preliminary research posted on the project's website. Developers and local governments in the following cities have expressed serious interest in establishing a vertical farm: Incheon (South Korea), Abu Dhabi (United Arab Emirates), and Dongtan (China), New York City, Portland, Ore., Los Angeles, Las Vegas, Seattle, Surrey, B.C., Toronto, Paris, Bangalore, Dubai, Shanghai and Beijing. The Illinois Institute of Technology is now crafting a detailed plan for Chicago. It is suggested that prototype versions of vertical farms should be created first, possibly at large universities interested in the research of vertical farms, in order to prevent failures such as the Biosphere 2 project in Oracle, Arizona.

In 2009, the world's first pilot production system was installed at Paignton Zoo Environmental Park in the United Kingdom. The project showcased a technological solution for vertical farming and provided a physical base to conduct research into sustainable urban food production. The produce is used to feed the zoo's animals while the project enables evaluation of the systems and provides an educational resource to advocate for change in unsustainable land use practices that impact upon global biodiversity and ecosystem services,

In 2010, the Green Zionist Alliance proposed a resolution at the 36th World Zionist Congress calling on Keren Kayemet L'Yisrael (Jewish National Fund in Israel) to develop vertical farms in Israel.

In 2012, the world's first commercial vertical farm was opened in Singapore, developed by Sky Greens Farms, and is three stories high. They currently have over 100 towers that stand at nine meters tall.

In 2013 (July 18) the Association for Vertical Farming (AVF) was founded in Munich (Germany). By May 2015 the AVF already expanded with regional chapters all over Europe, Asia, USA, Canada and the United Kingdom. This internationally active non-profit organization unites growers and inventors to improve food security and the sustainable development of Vertical Farming. The AVF also focuses on advancing vertical farming technologies, designs and businesses by hosting international info-days, workshops and summits. They developed a glossary to bring consistency to the industry and plan on helping to standardize the technologies.

LANDSCAPING

Landscaping requires expertise in horticulture and artistic design.Landscaping refers to any activity that modifies the visible features of an area of land, including:

- living elements, such as flora or fauna; or what is commonly called gardening.
- natural elements such as landforms, terrain shape and elevation, or bodies of water; and
- abstract elements such as the weather and lighting conditions.

Understanding the Land

Construction requires study and observation. It is not the same in different parts of the world. Landscaping varies according to different regions.Therefore, normally local natural experts are recommended if it is done for the first time. Understanding of the site is one of the chief essentials for successful landscaping. Different natural features like terrain, topography, soil qualities, prevailing winds, depth of the frost line, and the system of native flora and fauna must be taken into account. Sometimes the land is not fit for landscaping. In order to landscape it, the land must be reshaped. This reshaping of land is called grading.

Removal of earth from the land is called cutting while when earth is added to the slope, it is called filling. Sometimes the grading process may involve removal of excessive waste (landfills), soil and rocks, which the designers should take into account while in the planning stage.

Tools

First the landscaping contractor makes a letter which is a rough design and layout of what could be done with the land in order to achieve the desired outcome. Different pencils are required to make graphics of the picture. Landscaping has become more technological than natural, as few projects begin without bulldozers, lawnmowers, or chainsaws. Different areas have different qualities of plants. Fertilizers are required for this purpose in excess amounts as natural landscaping is done. Some landscapers prefer to use mix gravel with rocks of varying sizes to add interest in large areas.

REFERENCES

1. hortus. Charlton T. Lewis and Charles Short. A Latin Dictionary on Perseus Project.
2. Harper, Douglas. "horticulture". Online Etymology Dictionary.
3. Entry for yard Dictionary.com (presenting information supposedly from Random House Dictionary)

4. "Archived copy". Archived from the original on September 10, 2012. Retrieved September 21, 2012.

5. Fullagar, Richard, Judith Field, Tim Denham, and Carol Lentfer (2006) Early and mid Holocene tool-use and processing of taro (Colocasia esculenta), yam (Dioscorea sp.) and other plants at Kuk Swamp in the highlands of Papua New Guinea Journal of Archaeological Science 33: 595–614

6. von Hagen, V.W. (1957) The Ancient Sun Kingdoms Of The Americas. Ohio: The World Publishing Company

7. Solomon, Dawit, Johannes Lehmann, Janice Thies, Thorsten Schafer, Biqing Liang, James Kinyangi, Eduardo Neves, James Petersen, Flavio Luizao, and Jan Skjemstad, Molecular signature and sources of biochemical recalcitrance of organic carbone in Amazonian Dark Earths, Geochemica et cosmochemica ACTA 71.9 2285-2286 (2007) ("Amazonian Dark Earths (ADE) are a unique type of soils apparently developed between 500 and 9000 years B.P. through intense anthropogenic activities such as biomass-burning and high-intensity nutrient depositions on pre-Columbian Amerindian settlements that transformed the original soils into Fimic Anthrosols throughout the Brazilian Amazon Basin.") (internal citations omitted)

8. Glaser, Bruno, Johannes Lehmann, and Wolfgang Zech, Ameliorating physical and chemical properties of highly weathered soils in the tropics with charcoal – a review, Biology and Fertility of Soils 35.4 219-220 (2002) ("These so called Terra Preta do Indio (Terra Preta) characterize the settlements of pre-Columbian Indios. In Terra Preta soils large amounts of black C indicate a high and prolonged input of carbonized organic matter probably due to the production of charcoal in hearths, whereas only low amounts of charcoal are added to soils as a result of forest fires and slash-and-burn techniques.") (internal citations omitted)

9. McGee, J.R. and Kruse, M. (1986) Swidden horticulture among the Lacandon Maya [videorecording (29 mins.)]. University of California, Berkeley: Extension Media Center

10. Thompson, S.I. (1977) Women, Horticulture, and Society in Tropical America. American Anthropologist, N.S., 79: 908–10

11. Douglas John McConnell (2003). The Forest Farms of Kandy: And Other Gardens of Complete Design. p. 1. ISBN 9780754609582.

12. Douglas John McConnell (1992). The forest-garden farms of Kandy, Sri Lanka. p. 1. ISBN 9789251028988.

13. "A Brief History of Gardening". Retrieved 2010-06-04.

14. Ryrie, Charlie (2004). The Cottage Garden: How to Plan and Plant a Garden That Grows Itself. Collins & Brown. p. 7. ISBN 1-84340-216-5.

15. Scott-James, Anne; Osbert Lancaster (2004). The Pleasure Garden: An Illustrated History of British Gardening. Frances Lincoln Publishers. p. 80. ISBN 978-0-7112-2360-8.

16. Anne Scott-James, The Cottage Garden (London: Lane) 1981, de-mythologised the origins of the English cottage garden, and its treasured topiary among the vegetables and flowers, popularly supposed to represent heirlooms from the seventeenth century.
17. A Biographical Dictionary of British Architects, 1600–1840, Howard Colvin, Yale University Press, 2008 ISBN 0-300-12508-9, p 659
18. Lloyd, Christopher; Richard Bird (1999). The Cottage Garden. Jacqui Hurst. Dorling Kindersley. pp. 6–9. ISBN 978-0-7513-0702-3.
19. Yves-Marie Allain and Janine Christiany, L'Art des jardins en Europe, Citadelles and Mazenod, Paris, 2006.
20. Boults, Elizabeth and Chip Sullivan (2010). Illustrated History of Landscape Design. John Wiley and Sons. p. 175. ISBN 0-470-28933-3.
21. ^ Jump up to:a b http://www.richsoil.com/hugelkultur/
22. Hemenway, Toby (2009). Gaia's Garden: A Guide to Home-Scale Permaculture. Chelsea Green Publishing. pp. 84-85. ISBN 978-1-60358-029-8.
23. http://permaculture.org.au/2009/12/11/greening-the-desert-ii-final/
24. "What is a community garden?". American Community Garden Association. 2007.
25. Hannah, A. K. & Oh, P. (2000) Rethinking Urban Poverty: A look at Community Gardens. Bulletin of Science, Technology and & Society. 20(3). 207-216.
26. Ferris, J., Norman, C. & Sempik, J. (2001) People, Land and Sustainability: Community Gardens and the Social Dimension of Sustainable Development. Social Policy and Administration. 35(5). 559-568.
27. Meet the urban sharecroppers The Guardian, 4 September 2008
28. "A Common Pesticide Decreases Foraging Success and Survival in Honey Bees". Retrieved 2012-03-29.
29. "Mole-ested". Retrieved 2014-05-28.
30. "The Self-Sufficient Gardener Podcast--Episode 24 Companion Planting and Crop Rotation". Retrieved 2010-08-13.
31. Hix, John. 1974. The glass house. Cambridge, Mass: MIT Press.
32. Pati, Ranjan; Abelar, Michael (27 May 2015). "The Application and Optimization of Metal Reflectors to Vertical Greenhouses to Increase Plant Growth and Health". Journal of Agricultural Engineering and Biotechnology: 63–71. doi:10.18005/JAEB0302003.
33. "Glossary for Vertical Farming". Retrieved 2016-01-06.
34. Vertical farming (1915). Wilmington, Del.: E. I. duPont de Nemours Powder Co. Retrieved 2011-07-23.
35. Venkataraman, Bina (2008-07-15). "Country, the City Version: Farms in the Sky Gain New Interest". The New York Times. Retrieved 2011-01-05.
36. https://www.youtube.com/watch?v=1t-PUIXUPgw
37. https://www.youtube.com/watch?v=zWommg6JVfY

38. https://www.youtube.com/watch?v=r-Iymp-gdbo
39. https://www.youtube.com/watch?v=PrLTuGKcvPA
40. "The Rise of Urban Farming". Smithsonian Magazine. Retrieved 2011-01-05.
41. Douglas, James Sholto (1975). Hydroponics: The Bengal System(5th ed.). New Dehli: Oxford University Press. ISBN 9780195605662.
42. Nelson, Bryn (2007-12-12). "Could vertical farming be the future? Nelson, B. (2008).". MSNBC. Retrieved 2010-11-10.
43. George Monbiot (2010-08-16). "Monbiot, G. (16 August 2010). Greens living in ivory towers now want to farm them too. 'The". London: Guardian. Retrieved 2010-11-10.
44. https://store.ctbuh.org/PDF_Previews/Journal/CTBUHJournal_2009-2.pdf
45. Ruby, Andreas. Groundscapes pp. 87–93
46. Eaton, Ruth. Ideal Cities p. 239
47. link Archived August 11, 2011, at the Wayback Machine.
48. http://www.impactlab.net/wp-content/uploads/2008/10/editt-tower-by-tr-hamzah-and-yeang.jpg
49. http://ps-survival.com/PS/Hydroponics/Hydroponics_The_Bengal_System_1959.pdf
50. Arnie Cooper (May 19, 2009). "Going up? Farming in High Rises Raises Hopes". Miller-mccune.com. Retrieved 2010-11-10.
51. Whyte, Murray (2008-07-27). "Is high rise farming in Toronto's future?". Toronto Star. Retrieved 2008-08-12.
52. "Sky Farm Proposed for Downtown Toronto". TreeHugger. Retrieved 2009-03-14.
53. Venkataraman, Bina (2008-07-15). "Country, the City Version: Farms in the Sky Gain New Interest". The New York Times.
54. Shute, Nancy (2007-05-20). "Farm of the Future? Someday food may grow in skyscrapers". U.S. News & World Report.
55. Feldman, Amy (2007-07-11). "Skyscraper Farms". Popular Science.
56. Despommier, Dickson (November 2009). "The Rise of Vertical Farms". Scientific American. New york: Scientific American Inc. 301 (5): 60–67. ISSN 0036-8733.
57. http://inhabitat.com/vertical-harvest-of-jackson-hole-will-grow-organic-produce-even-in-the-winter/
58. "Vertical Fresh Farms LLC, Buffalo, NY | StateLog". www.statelog.com. Retrieved 2015-11-05.
59. Clive Clifton says: (2009-08-24). "Vertical Farming: Too Far Outside the Box? |". E4capital.com. Retrieved 2010-11-10.
60. Michael Bomford says: (2010-12-02). "Energy and VerticalFarms". Retrieved 2011-03-08.

61. Evans, P. (July 22, 2009). Local food no green panacea: professor. CBC News http://www.cbc.ca/consumer/story/2009/07/22/consumer-local-food.html
62. https://www.youtube.com/watch?v=KZN3ucwePGk
63. "Does it really stack up?". The Economist. 2010-12-09.
64. "Starting a Commercial Greenhouse Business". Omafra.gov.on.ca. 2003-07-28. Retrieved 2010-11-10.
65. Pocket World in Figures, The Economist, 2011 ed. pg 64
66. Roach, J. (June 30, 2009). "High-Rise Farms: The Future of Food?". National Geographic News.
67. "George Monbiot – Towering Lunacy". Monbiot.com. Retrieved 2013-06-12.
68. The Economist (December 9, 2010). "Vertical farming: Does it really stack up?".
69. "Crops | Greenhouse | Greenhouse Energy Calculations | Manitoba Agriculture, Food and Rural Initiatives | Province of Manitoba". Gov.mb.ca. Retrieved 2010-11-10.
70. "About The Plant". The Plant.
71. "Archived copy" (PDF). Archived from the original (PDF) on 2011-07-19. Retrieved 2010-11-09.
72. Blom, T.J.; W.A. Straver; F.J. Ingratta; Shalin Khosla; Wayne Brown (December 2002). "Carbon Dioxide In Greenhouses". Retrieved 2010-10-10.
73. Light pollution handbook – Google Books. Books.google.ca. 2004. ISBN 978-1-4020-2665-2. Retrieved 2010-11-10.
74. "Treating and Recycling Irrigation Runoff". Aggie-horticulture.tamu.edu. Retrieved 2013-06-12.
75. Despommier, D. (2008). "Vertical Farm Essay I". Vertical Farm. Retrieved 2009-06-26.
76. The Vertical Farm Project. 2009. "Agriculture for the 21st Century and Beyond". http://www.verticalfarm.com/
77. Frediani, K. L. 2010 Feeding time at the Zoo. The Horticulturalist. April 2010, pp.12–15 Institute of Horticulture.
78. Frediani, K. L. (2011), "High rise food", The Horticulturalist, October, pp.18–20.
79. Despommier, D. (2008). "Vertical Farm Essay I". Vertical Farm. Retrieved 2009-06-26.
80. "Vertical Farm Video". Discovery Channel. 2009-04-23. Retrieved 2009-06-26.
81. "Dwarf Wheat grown aboard the International Space Station". NASA. 9 February 2003. Retrieved 17 November 2009.
82. Michael Pollan (September 9, 2009). "Big Food vs. Big Insurance". The New York Times.
83. "A Farm on Every Floor", The New York Times, August 23, 2009

84. Vertical take off, Fresh Produce Journal, 28 January 2011
85. S.L. Davis (2001). "The least harm principle suggests that humans should eat beef, lamb, dairy, not a vegan diet". Proceedings of the Third Congress of the European Society for Agricultural and Food Ethics. pp. 449–450.
86. Navarro, L.M. & Pereira, H.M. (2012). "Rewilding Abandoned Landscapes in Europe.": 900–912.
87. "Food and Culture Spring 2011 course". ocw.mit.edu.
88. Case Study — Landfill Power Generation, H. Scott Matthews, Green Design Initiative, Carnegie Mellon University. "Archived copy" (PDF). Archived from the original (PDF) on 2008-12-03. Retrieved 2009-02-09. Retrieved 07.02.09
89. Folke Günther (2013-01-06). "The folkewall, greywater purification AND vertical growing". Holon.se. Retrieved 2013-06-12.
90. McConnell, Kathryn (2008-07-01). "Vertical Farms Grow Food by Growing Up, Not Out". Bureau of International Information Programs. United States Department of State. Retrieved 2008-08-12.
91. "Las Vegas to Build World's First 30 Story Vertical Farm". Nextenergynews.com. Retrieved 2013-06-12.
92. Despommier, Dickson (2010-06-15). "Dickson Despommier. November 2009. "Growing Skyscrapers: The Rise of Vertical Farms."". Scientificamerican.com. Retrieved 2010-11-10.
93. "Vertical plant production as a public exhibit at Paignton Zoo" (PDF). Proceedings of the 4th World Botanic Garden Congress, Botanic Gardens Conservation International. June 2010. Retrieved 2013-01-27.
94. "Green Zionist Alliance (GZA) – Bold Resolutions for 36th World Zionist Congress".
95. "World's first commercial vertical farm opens in Singapore". Io9.com. Retrieved 2013-06-12.
96. "Urban farming looking up in Singapore – CNN.com". Edition.cnn.com. Retrieved 2013-06-12.
97. "vertical farming infoday munich - agritecture.com". Retrieved 2015-06-01.
98. "Vertical Farming technologies trends - agritecture.com". Retrieved 2015-06-01.
99. Natural Landscaping: Designing With Native Plant Communities - John Diekelmann, Robert M. Schuster - Google Books. Books.google.co.uk. Retrieved 2013-04-10.
100. Landscaping Principles and Practices - Jack Ingels - Google Books. Books.google.co.uk. 2009-01-15. Retrieved 2013-04-10.
101. Landscaping - William Slack - Google Books. Books.google.co.uk. Retrieved 2013-04-10.
102. Taylor's Master Guide to Landscaping - Rita Buchanan - Google Books. Books.google.co.uk. Retrieved 2013-04-10.

103. John Smith. “Landscaping by landscape gardeners: Methods and Tactics”. New Ways Landscaping Design. Retrieved 2016-06-14.

104. Sharon Cohoon and Jim McCausland. “How to Landscape Gravel - Page 2”. Sunset.com. Archived from the original on March 19, 2012. Retrieved 2013-04-10.

CHAPTER 2

Floriculture

Chapter Contents

- **Introduction**
- **Plant Breeding**
- **Gartons Agricultural Plant Breeders**
- **Norman Borlaug**

Floriculture is the growing of cut flowers, potted flowering and foliage plants, and bedding plants in greenhouses and/or in fields. There are several thousand different species of flowers and plants that are grown as commercial crops.

INTRODUCTION

Floriculture, or flower farming, is a discipline of horticulture concerned with the cultivation of flowering and ornamental plants for gardens and for floristry, comprising the floral industry. The development, via plant breeding, of new varieties is a major occupation of floriculturists.

Floriculture crops include bedding plants, houseplants, flowering garden and pot plants, cut cultivated greens, and cut flowers. As distinguished from nursery crops, floriculture crops are generally herbaceous. Bedding and garden plants consist of young flowering plants (annuals and perennials) and vegetable plants. They are grown in cell packs (in flats or trays), in pots, or in hanging baskets, usually inside a controlled environment, and sold largely for gardens and landscaping. Pelargonium ("geraniums"), Impatiens ("busy lizzies"), and Petunia are the best-selling bedding plants. The many cultivars of Chrysanthemum are the major perennial garden plant in the United States.

Flowering plants are largely sold in pots for indoor use. The major flowering plants are poinsettias, orchids, florist chrysanthemums, and finished florist azaleas. Foliage plants are also sold in pots and hanging baskets for indoor and patio use, including larger specimens for office, hotel, and restaurant interiors.

Cut flowers are usually sold in bunches or as bouquets with cut foliage. The production of cut flowers is specifically known as the cut flower industry. Farming flowers and foliage employs special aspects of floriculture, such as spacing, training and pruning plants for optimal flower harvest; and post-harvest treatment such as chemical treatments, storage, preservation and packaging. In Australia and the United States some species are harvested from the wild for the cut flower market.

A retail greenhouse shows some of the diversity of floricultural plants

PLANT BREEDING

Plant breeding is the art and science of changing the traits of plants in order to produce desired characteristics. Plant breeding can be accomplished through many different techniques ranging from simply selecting plants with desirable characteristics for propagation, to more complex molecular techniques (see cultigen and cultivar).

Plant breeding has been practiced for thousands of years, since near the beginning of human civilization. It is practiced worldwide by individuals such as gardeners and farmers, or by professional plant breeders employed by organizations such as government institutions, universities, crop-specific industry associations or research centers.

International development nation agencies believe that breeding new crops is important for ensuring sustainability by developing new varieties that are higher-yielding, disease resistant, drought-resistant or regionally adapted to different environments and growing conditions.

The Yecoro wheat (right) cultivar is sensitive to salinity, plants resulting from a hybrid cross with cultivar W4910 (left) show greater tolerance to high salinity

Plant breeding started with sedentary agriculture and particularly the domestication of the first agricultural plants, a practice which is estimated to date back 9,000 to 11,000 years. Initially early farmers simply selected food plants with particular desirable characteristics, and employed these as progenitors for subsequent generations, resulting in an accumulation of valuable traits over time.

Gregor Mendel's experiments with plant hybridization led to his establishing laws of inheritance. Once this work became well known, it formed the basis of the new science of genetics, which stimulated research by many plant scientists dedicated to improving crop production through plant breeding.

Modern plant breeding is applied genetics, but its scientific basis is broader, covering molecular biology, cytology, systematics, physiology, pathology, entomology, chemistry, and statistics (biometrics). It has also developed its own technology.

Classical Plant Breeding

One major technique of plant breeding is selection, the process of selectively propagating plants with desirable characteristics and eliminating or "culling" those with less desirable characteristics.

Another technique is the deliberate interbreeding (crossing) of closely or distantly related individuals to produce new crop varieties or lines with desirable properties. Plants arc crossbred to introduce traits/genes from one variety or line into a new genetic background. For example, a mildew-resistant pea may be crossed with a high-yielding but susceptible pea, the goal of the cross being to introduce mildew resistance without losing the high-yield characteristics. Progeny from the cross would then be crossed with the high-yielding parent to ensure that the progeny were most like the high-yielding parent, (backcrossing). The progeny from that cross would then be tested for yield (selection, as described above) and mildew resistance and high-yielding resistant plants would be further developed. Plants may also be crossed with themselves to produce inbred varieties for breeding. Pollinators may be excluded through the use of pollination bags.

Classical breeding relies largely on homologous recombination between chromosomes to generate genetic diversity. The classical plant breeder may also make use of a number of *in vitro* techniques such as protoplast fusion, embryo rescue or mutagenesis (see below) to generate diversity and produce hybrid plants that would not exist in nature.

Traits that breeders have tried to incorporate into crop plants include:

- Improved quality, such as increased nutrition, improved flavor, or greater beauty
- Increased yield of the crop
- Increased tolerance of environmental pressures (salinity, extreme temperature, drought)
- Resistance to viruses, fungi and bacteria
- Increased tolerance to insect pests
- Increased tolerance of herbicides
- Longer storage period for the harvested crop

Before World War II

Garton's catalogue from 1902

Successful commercial plant breeding concerns were founded from the late 19th century. Gartons Agricultural Plant Breeders in England was established in the 1890s by John Garton, who was one of the first to commercialize new varieties of agricultural crops created through cross-pollination. The firm's first introduction was Abundance Oat, one of the first agricultural grain varieties bred from a *controlled* cross, introduced to commerce in 1892. In the early 20th century, plant breeders realized that Mendel›s findings on the non-random nature of inheritance could be applied to seedling populations produced through deliberate pollinations to predict the frequencies of different types. Wheat hybrids were bred to increase the crop production of Italy during the so-called

«Battle for Grain» (1925–1940). Heterosis was explained by George Harrison Shull. It describes the tendency of the progeny of a specific cross to outperform both parents. The detection of the usefulness of heterosis for plant breeding has led to the development of inbred lines that reveal a heterotic yield advantage when they are crossed. Maize was the first species where heterosis was widely used to produce hybrids.

Statistical methods were also developed to analyze gene action and distinguish heritable variation from variation caused by environment. In 1933 another important breeding technique, cytoplasmic male sterility (CMS), developed in maize, was described by Marcus Morton Rhoades. CMS is a maternally inherited trait that makes the plant produce sterile pollen. This enables the production of hybrids without the need for labor-intensive detasseling.

These early breeding techniques resulted in large yield increase in the United States in the early 20th century. Similar yield increases were not produced elsewhere until after World War II, the Green Revolution increased crop production in the developing world in the 1960s.

After World War II

In vitro-culture of Vitis (grapevine), Geisenheim Grape Breeding Institute

Following World War II a number of techniques were developed that allowed plant breeders to hybridize distantly related species, and artificially induce genetic diversity.

When distantly related species are crossed, plant breeders make use of a number of plant tissue culture techniques to produce progeny from otherwise fruitless mating. Interspecific and intergeneric hybrids are produced from a cross of related species or genera that do not normally sexually reproduce with each other. These crosses are referred to as *Wide crosses*. For example, the cereal triticale is a wheat and rye hybrid. The cells in the plants derived from the first generation created from the cross contained an uneven number of chromosomes and as result was sterile. The cell division inhibitor colchicine was used to double the number of chromosomes in the cell and thus allow the production of a fertile line.

Failure to produce a hybrid may be due to pre- or post-fertilization incompatibility. If fertilization is possible between two species or genera, the hybrid embryo may abort before maturation. If this does occur the embryo resulting from an interspecific or intergeneric cross can sometimes be rescued and cultured to produce a whole plant. Such a method is referred to as Embryo Rescue.

Hybrids may also be produced by a technique called protoplast fusion. In this case protoplasts are fused, usually in an electric field. Viable recombinants can be regenerated in culture.

Chemical mutagens like EMS and DMS, radiation and transposons are used to generate mutants with desirable traits to be bred with other cultivars - a process known as *Mutation Breeding*. Classical plant breeders also generate genetic diversity within a species by exploiting a process called somaclonal variation, which occurs in plants produced from tissue culture, particularly plants derived from callus. Induced polyploidy, and the addition or removal of chromosomes using a technique called chromosome engineering may also be used.

When a desirable trait has been bred into a species, a number of crosses to the favored parent are made to make the new plant as similar to the favored parent as possible. Returning to the example of the mildew resistant pea being crossed with a high-yielding but susceptible pea, to make the mildew resistant progeny of the cross most like the high-yielding parent, the progeny will be crossed back to that parent for several generations (See backcrossing). This process removes most of the genetic contribution of the mildew resistant parent. Classical breeding is therefore a cyclical process.

With classical breeding techniques, the breeder does not know exactly what genes have been introduced to the new cultivars. Some scientists therefore argue that plants produced by classical breeding methods should undergo the same safety testing regime as genetically modified plants. There have been instances where plants bred using classical techniques have been unsuitable for human consumption, for example the poisonsolanine was unintentionally increased to unacceptable levels in certain varieties of potato through plant breeding. New potato varieties are often screened for solanine levels before reaching the marketplace.

Modern Plant Breeding

Modern plant breeding may use techniques of molecular biology to select, or in the case of genetic modification, to insert, desirable traits into plants. Application of biotechnology or molecular biology is also known as molecular breeding (see: Molecular breeding).

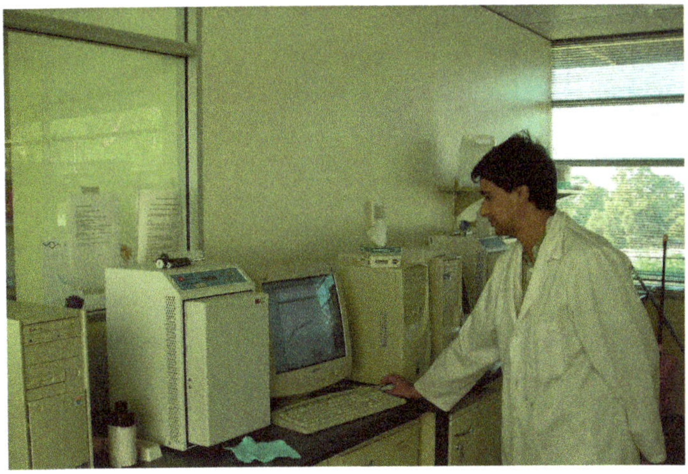

Modern facilities in molecular biology have converted classical plant breeding to molecular plant breeding

Steps of Plant Breeding

The following are the major activities of plant breeding:

- Collection of variation
- Selection
- Evaluation
- Release
- Multiplication
- Distribution of the new variety
- Selling to people

Marker Assisted Selection

Sometimes many different genes can influence a desirable trait in plant breeding. The use of tools such as molecular markers or DNA fingerprinting can map thousands of genes. This allows plant breeders to screen large populations of plants for those that possess the trait of interest. The screening is based on the presence or absence of a certain gene as determined by laboratory procedures, rather than on the visual identification of the expressed trait in the plant.

Reverse Breeding and Doubled Haploids (DH)

A method for efficiently producing homozygous plants from a heterozygous starting plant, which has all desirable traits. This starting plant is induced to produce doubled haploid from haploid cells, and later on creating homozygous/doubled haploid plants from those cells. While in natural offspring genetic recombination occurs and traits can be unlinked from each other, in doubled haploid cells and in the resulting DH plants recombination is no longer an issue. There, a recombination between two corresponding chromosomes does not lead to un-linkage of alleles or traits, since it just leads to recombination with its identical copy. Thus, traits on one chromosome stay linked. Selecting those offspring having the desired set of chromosomes and crossing them will result in a final F1 hybrid plant, having exactly the same set of chromosomes, genes and traits as the starting hybrid plant. The homozygous parental lines can reconstitute the original heterozygous plant by crossing, if desired even in a large quantity. An individual heterozygous plant can be converted into a heterozygous variety (F1 hybrid) without the necessity of vegetative propagation but as the result of the cross of two homozygous/doubled haploid lines derived from the originally selected plant. patent

Genetic Modification

Genetic modification of plants is achieved by adding a specific gene or genes to a plant, or by knocking down a gene with RNAi, to produce a desirable phenotype. The plants resulting from adding a gene are often referred to as transgenic plants. If for genetic modification genes of the species or of a crossable plant are used under control of their native promoter, then they are called cisgenic plants. Sometimes genetic modification can produce a plant with the desired trait or traits faster than classical breeding because the majority of the plant›s genome is not altered.

To genetically modify a plant, a genetic construct must be designed so that the gene to be added or removed will be expressed by the plant. To do this, a promoter to drive transcription and a termination sequence to stop transcription of the new gene, and the gene or genes of interest must be introduced to the plant. A marker for the selection of transformed plants is also included. In the laboratory, antibiotic resistance is a commonly used marker: Plants that have been successfully transformed will grow on media containing antibiotics; plants that have not been transformed will die. In some instances markers for selection are removed by backcrossing with the parent plant prior to commercial release.

The construct can be inserted in the plant genome by genetic recombination using the bacteria *Agrobacterium tumefaciens* or *A. rhizogenes*, or by direct methods like the gene gun or microinjection. Using plant viruses to insert genetic constructs into plants is also a possibility, but the technique is limited by the host range of the virus. For example, Cauliflower mosaic virus (CaMV) only infects cauliflower and related species. Another limitation of viral vectors is that the virus is not usually passed on the progeny, so every plant has to be inoculated.

The majority of commercially released transgenic plants are currently limited to plants that have introduced resistance to insect pests and herbicides. Insect resistance is achieved through incorporation of a gene from *Bacillus thuringiensis* (Bt) that encodes a protein that is toxic to some insects. For example, the cotton bollworm, a common cotton pest, feeds on Bt cotton it will ingest the toxin and die. Herbicides usually work by binding to certain plant enzymes and inhibiting their action. The enzymes that the herbicide inhibits are known as the herbicides *target site*. Herbicide resistance can be engineered into crops by expressing a version of *target site* protein that is not inhibited by the herbicide. This is the method used to produce glyphosate resistant crop plants (See Glyphosate)

Genetic modification of plants that can produce pharmaceuticals (and industrial chemicals), sometimes called *pharming*, is a rather radical new area of plant breeding.

Issues and Concerns

Modern plant breeding, whether classical or through genetic engineering, comes with issues of concern, particularly with regard to food crops. The question of whether breeding can have a negative effect on nutritional value is central in this respect. Although relatively little direct research in this area has been done, there are scientific indications that, by favoring certain aspects of a plant's development, other aspects may be retarded. A study published in the *Journal of the American College of Nutrition* in 2004, entitled *Changes in USDA Food Composition Data for 43 Garden Crops, 1950 to 1999*, compared nutritional analysis of vegetables done in 1950 and in 1999, and found substantial decreases in six of 13 nutrients measured, including 6% of protein and 38% of riboflavin. Reductions in calcium, phosphorus, iron and ascorbic acid were also found. The study, conducted at the Biochemical Institute, University of Texas at Austin, concluded in summary: *"We suggest that any real declines are generally most easily explained by changes in cultivated varieties between 1950 and 1999, in which there may be trade-offs between yield and nutrient content."*

The debate surrounding genetically modified food during the 1990s peaked in 1999 in terms of media coverage and risk perception, and continues today - for example, "*Germany has thrown its weight behind a growing European mutiny over genetically modified crops by banning the planting of a widely grown pest-resistant corn variety.*" The debate encompasses the ecological impact of genetically modified plants, the safety of genetically modified food and concepts used for safety evaluation like substantial equivalcncc. Such concerns are not new to plant breeding. Most countries have regulatory processes in place to help ensure that new crop varieties entering the marketplace are both safe and meet farmers' needs. Examples include variety registration, seed schemes, regulatory authorizations for GM plants, etc.

Plant breeders' rights is also a major and controversial issue. Today, production of new varieties is dominated by commercial plant breeders, who seek to protect their work and collect royalties through national and international agreements based in intellectual property rights. The range of related issues is complex. In the simplest terms, critics of the increasingly restrictive regulations argue that, through a combination of technical and economic pressures, commercial breeders are reducing biodiversity and significantly constraining individuals (such as farmers) from developing and trading seed on a regional level. Efforts to strengthen breeders' rights, for example, by lengthening periods of variety protcction, are ongoing.

When new plant breeds or cultivars are bred, they must be maintained and propagated. Some plants are propagated by asexual means while others are propagated by seeds. Seed propagated cultivars require specific control over seed source and production procedures to maintain the integrity of the plant breeds results. Isolation is necessary to prevent cross contamination with related plants or the mixing of seeds after harvesting. Isolation is normally accomplished by planting distance but in certain crops, plants are enclosed in greenhouses or cages (most commonly used when producing F1 hybrids.)

Role of Plant Breeding in Organic Agriculture

Critics of organic agriculture claim it is too low-yielding to be a viable alternative to conventional agriculture. However, part of that poor performance may be the result of growing poorly adapted varieties. It is estimated that over 95% of organic agriculture is based on conventionally adapted varieties, even though the production environments found in organic vs. conventional farming systems are vastly different due to their distinctive management practices. Most notably, organic farmers have fewer inputs available than conventional growers to control their production environments. Breeding varieties specifically adapted to the unique conditions of organic agriculture is critical for this sector to realize its full potential. This requires selection for traits such as:

- Water use efficiency
- Nutrient use efficiency (particularly nitrogen and phosphorus)
- Weed competitiveness
- Tolerance of mechanical weed control
- Pest/disease resistance
- Early maturity (as a mechanism for avoidance of particular stresses)
- Abiotic stress tolerance (i.e. drought, salinity, etc...)

Currently, few breeding programs are directed at organic agriculture and until recently those that did address this sector have generally relied on indirect selection (i.e. selection in conventional environments for traits considered important for organic agriculture). However, because the difference between organic and conventional environments is large, a given genotype may perform very differently in each environment due to an interaction between genes and the environment (see gene-environment interaction). If this interaction is severe enough, an important trait required for the organic environment may not be revealed in the conventional environment, which can result in the selection of poorly adapted individuals. To ensure the most adapted varieties are identified, advocates of organic breeding now promote the use of direct selection (i.e. selection in the target environment) for many agronomic traits.

There are many classical and modern breeding techniques that can be utilized for crop improvement in organic agriculture despite the ban on genetically modified organisms. For instance, controlled crosses between individuals allow desirable genetic variation to be recombined and transferred to seed progeny via natural processes. Marker assisted selection can also be employed as a diagnostics tool to facilitate selection of progeny who possess the desired trait(s), greatly speeding up the breeding process. This technique has proven particularly useful for the introgression of resistance genes into new backgrounds, as well as the efficient selection of many resistance genes pyramided into a single individual. Unfortunately, molecular markers are not currently available for many important traits, especially complex ones controlled by many genes.

Addressing Global Food Security through Plant Breeding

For future agriculture to thrive there are necessary changes which must be made in accordance to arising global issues. These issues are arable land, harsh cropping conditions and food security which involves, being able to provide the world population with food containing sufficient nutrients. These crops need to be able to mature in several environments allowing for worldwide access, this is involves issues such as drought tolerance. These global issues are achievable through the process of plant breeding, as it offers the ability to select specific genes allowing the crop to perform at a level which yields the desired results.

Minimal Land Degradation

Land degradation is a major issue, as it can negatively impact the capability of the land to be productive. Poor agricultural management has a huge impact on the degradation of soil worldwide and it is Africa and Asia that are most affected. Through education and development of modified plants, these statistics can be reduced and agricultural land can become more productive. Plant breeding allows for an increase in yield with out the extra strain on the land. The genetically modified, Bt white maize, was introduced to South Africa and was surveyed in 33 large commercial farms and 368 small landholders properties and in both cases a higher yield was recorded.

Increased Yield without Expansion

With an increasing population, the production of food needs to increase with it. It is estimated that a 70% increase in food production is needed by 2050 in order to meet the Declaration of the World Summit on Food Security. But with the natural degradation of agricultural land, simply planting more crops is no longer a viable option. Therefore, new varieties of plants need to be developed through plant breeding that generates an increase of yield without relying on an increase in land area. An example of this can be seen in Asia, where food production per capita has increased twofold. This has been achieved through not only the use of fertilisers, but through the use of better crops that have been specifically designed for the area.

Breeding for Increased Nutritional Value

Plant breeding can contribute to global food security as it is a cost-effective tool for increasing nutritional value of forage and crops. Improvements in nutritional value for forage crops from the use of analytical chemistry and rumen fermentation technology have been recorded since 1960; this science and technology gave breeders the ability to screen thousands of samples within a small amount of time, meaning breeders could identify a high performing hybrid quicker. The main area genetic increases were made was in vitro dry matter digestibility (IVDMD) resulting in 0.7-2.5% increase, at just 1% increase in IVDMD a single Bos Taurus also known as beef cattle reported 3.2% increase in daily gains. This improvement indicates plant breeding is an essential tool in gearing future agriculture to perform at a more advanced level.

Breeding for Tolerance

Plant breeding of hybrid crops has become extremely popular worldwide in an effort to combat the harsh environment. With long periods of drought and lack of water or nitrogen stress tolerance has become a significant part of agriculture. Plant breeders have focused on identifying crops which will ensure crops perform under these conditions; a way to achieve this is finding strains of the crop that is resistance to drought conditions with low nitrogen. It is evident from this that plant breeding is vital for future agriculture to survive as it enables farmers to produce stress resistant crops hence improving food security.

Participatory Plant Breeding

Participatory plant breeding (PPB) is when farmers are involved in a crop improvement programme with opportunities to make decisions and contribute to the research process at different stages. Participatory approaches to crop improvement can also be applied when plant biotechnologies are being used for crop improvement.

List of Notable Plant Breeders

- Gartons Agricultural Plant Breeders
- Gregor Mendel
- Keith Downey
- Luther Burbank
- Nazareno Strampelli
- Niels Ebbesen Hansen
- Edger McFadden
- Norman Borlaug

GARTONS AGRICULTURAL PLANT BREEDERS

Dr John Garton, of the firm of Garton Brothers of Newton-le-Willows in the United Kingdom was the Originator of Scientific Farm Plant Breeding. He is credited as the first scientist to show that the common grain crops and many other plants are self-fertilizing. He also invented the process of multiple cross-fertilization of crop plants.

In 1898 the business became known as Gartons Limited and, under the inspired commercial leadership of George Peddie Miln, was to become the British Empire's largest plant breeding and seed company.

A public company from the start, its shares were traded on the London Stock Exchange from 1947.

John Garton

Commercial Start

Robert and John Garton made a commercial start as R. & J. Garton. They launched their first variety, 'Abundance' oat, in 1892.

Generous publicity followed in the press, together with the publication of articles by botanists in the Journal of the Royal Agricultural Society of England, and in the Transactions of the Royal Highland and Agricultural Society of Scotland in 1894 and 1898. Professor Robert Wallace (1853–1939) of the University of Edinburgh said 'Under the system originated by Mr John Garton an infinite number of new and distinct breeds of oats, barleys, wheats, clovers and grasses have been produced'.

In 1898 a public company was launched, Gartons Limited. It was based in Warrington. Many of the 600 or so subscribers for £50,000 cumulative preference shares of 6% rising to 10% were farmers.

George Peddie Miln joined the company as Managing Director, together with Robert Garton, Thomas R. Garton, Thomas Baxter and Arthur Smith as directors. Robert and John Garton agreed to continue to work for the Company for five years for £500 and to receive the entire ordinary share capital of the new company of £50,000.

It rapidly became the United Kingdom's best known plant breeding and seed company, and also exported seeds widely.

The Garton Lectureship at Edinburgh University

In 1900 an endowment was made to found the Garton Lectureship in Indian and Colonial Agriculture at the University of Edinburgh. The Garton Lectureship still exists as a biennial award to promising young lecturers in the School of Agriculture but is without emoluments and is no longer tied to colonial agriculture.

Publicity in USA

Noel Kingsbury Writes:

From the late nineteenth century on, seed companies began to play an increasingly important, if not dominant, role in breeding non-cereal crops and a major role in producing varieties for market gardening and for private growers. The production of new cereals was a somewhat different matter - the fact that they were so vitally important for national food supplies and involved large-scale and long-term work made it more likely that they would be the concern of government. There were exceptions though, one being the family firm of Gartons of Warrington, Lancashire, in the north of England. Their production of cereals - oats in particular - was appreciated as internationally important during the latter quarter of the nineteenth century and the early years of the twentieth; (The American wheat breeder) Mark Carleton visited them in 1898 and was reportedly astonished at their work, Garton varieties were widely exported throughout the British Empire (then by far and away the world's largest political unit) and the United States.

"That private companies could be so effective in breeding cereal grains indicated that there was no link of necessity between their improvement and the publicly funded research that was to so dominate this sector over so much of the next century."

In 1903 Professor Willet M. Hays (1859–1927) of the Agricultural Experiment Station in Minnesota, USA said 'No one has done more brilliant work in Agricultural Plant Breeding than Messrs. Garton, and this is from now on to be recognised.'

Attempted Intellectual Property Donation

The introduction to their 1899 Spring Catalogue reads:

Our original idea for the dissemination of the seed of these new breeds as the stocks became sufficiently large for the purpose, was through some public body as in the form of an annual free seed distribution upon similar lines to the free seed distributions carried out by the Governments of the United States, Canada, and several of the British Colonies.

On three successive occasions we approached Her Majesty's Government with this object in view, the first occasion being on the formation of the Board of Agriculture, in 1889, when we offered to hand over the whole of the valuable results, providing that body would undertake their dissemination and the continuance of the work, either in the form of an annual free seed distribution or at current market price.

Upon the last occasion our offer was accompanied by letters and reports from all the leading Agricultural Professors, Botanists, and Scientists in the Kingdom, setting forth the national benefit which would accrue from the dissemination of the results in the form we had suggested. The final reply of Her Majesty's Government, however, was that whilst fully recognising the value of the work, owing to there being no precedent upon which to act in such a matter, they were unable to avail themselves of the offer. This was much to be regretted for had our ideas been carried into effect the British farmer would have been placed in immediate possession of important results, which in the hands of a Public

Company would not reach him for many years. Our efforts in this direction not having been successful, and as we were not in a position to undertake the work of distribution ourselves, we have placed it in the lands of a Public Company, and we trust that the continued efforts made by us on behalf of the British farmer will be fully appreciated by him, through his support of the Company responsible for the distribution of the seed of our new breeds of agricultural plants

R. & J. Garton"

Crop Innovations

The firm's first historic introduction was 'Abundance' oat, the world's first agricultural plant variety bred from a *controlled* cross, introduced to commerce in 1892.

Among the other 170 crop varieties Gartons Limited bred and introduced to commerce were:

- ‹Standwell› barley, introduced in 1898, the first barley bred from a controlled cross.
- ‹White Monarch› wheat, introduced in 1899, the first wheat bred from a controlled cross.
- ‹Invincible› barley, introduced in 1899, the first crop plant control crossed for disease resistance.
- Perennialized red clover, introduced in 1898, the first controlled cross clover.
- Perennialized Italian ryegrass, introduced in 1907, the first controlled cross pasture grass.
- ‹Apex› winter wheat, introduced in 1967, the first wheat to be granted plant breeders› rights in the United Kingdom.

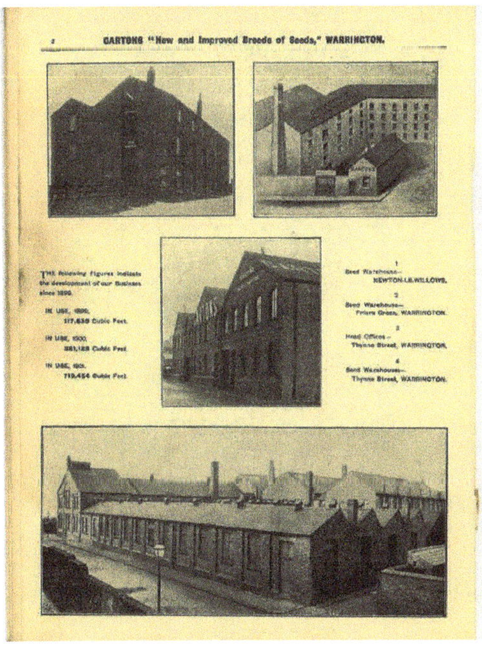

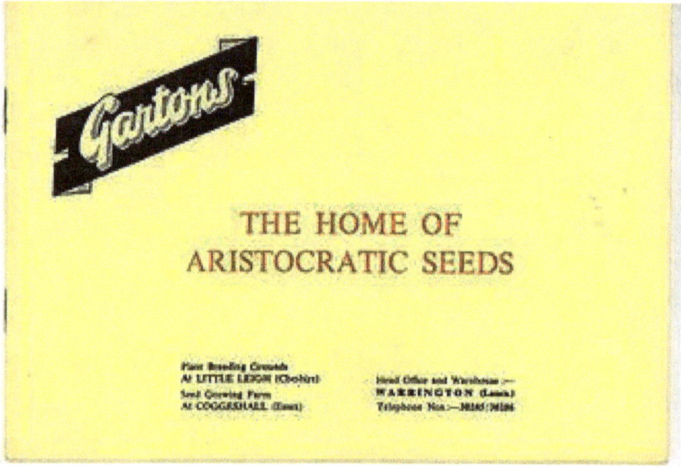
Gartons
THE HOME OF
ARISTOCRATIC SEEDS

Eight distinct grass strains
will be included in our mixture
for higher yield per acre
Gartons Own STRAINS

John Garton's honorary doctorate

On 23 March 1922 the Senatus Academicus of the University of Edinburgh offered to confer the honorary doctorate of LL.D on John Garton shortly before his death, which he duly accepted. At its meeting on 6 July 1922 the Senatus Academicus learned that John Garton had died.

The programme and report of the Graduation Ceremony held on 21 July 1922 reads

The Senatus Academicus recently conferred the Honorary Degree of Doctor of Laws upon the late John Garton, who duly accepted it. The Degree would have been formally conferred on the present occasion but for his lamented death. Mr Garton invented the process of multiple cross fertilisation of crop plants and has been the means in a great measure of revolutionising field culture by producing hundreds of new and improved varieties which have greatly increased the yields of all the common crops of the farm. The achievement proved to be of immense national importance during the War.

Mr Garton first showed that the common grain crops and many other crop plants are self-fertilising. Up till that time they were generally believed to be fertilised by wind or insects. Mr Garton's results got in crossing different species of grasses helped to develop the modern conception of species.

Twenty-two tears ago Mr Garton provided the means to establish the Garton Lectures on Colonial and Indian Agriculture, and subsequently he permanently endowed them as an integral part of the work of the Chair of Agriculture.

Plant Breeding Grounds

The plant breeding grounds were initially at Newton-le-Willows but moved to Acton Grange, two and a half miles south west of Warrington before settling in about 1930 at Little Leigh near Northwich in Cheshire. A seed development farm was located in Essex, and root crop trials were located on farms in the north of England and in Scotland. Traditionally groups of farmers were invited in mid-summer to inspect the plant breeding grounds and be entertained by the Company.

Initially the Seed Warehouse for cleaning and distributing seed was in Newton-le-Willows but moved to Friars Green in Warrington in 1899 by which time the offices were at Thynne Street, Warrington. A purpose built seven story Seed Warehouse and separate Head Office were built at Arpley, Warrington in 1910. There was an L. M. S. railway siding into the Seed Warehouse. On 25 April 1912 the Seed Warehouse burned down but quickly rebuilt largely by the same builders. Seed cleaning machinery, some unique to the Company, ensured the purity of the product. As time went by fewer seeds were 'picked' or cleaned by hand by upwards of one hundred staff as machinery became more sophisticated. Across the top of roof of the warehouse was the company's name which had to be disguised during wartime.

From the beginning Gartons Limited tested its seeds for purity and germination at its own seed testing laboratories in Warrington. The 1920 Seeds Act, for the first time, made testing and declaring for purity and germination a legal requirement for all seed companies. The Official Seed Testing Station was created in 1917, firstly in Victoria Street in Westminster, London and then in 1921 within the newly formed National Institute of Agricultural Botany in Cambridge. Larger seed companies including Gartons Limited were licensed to carry out their own purity and germination testing.

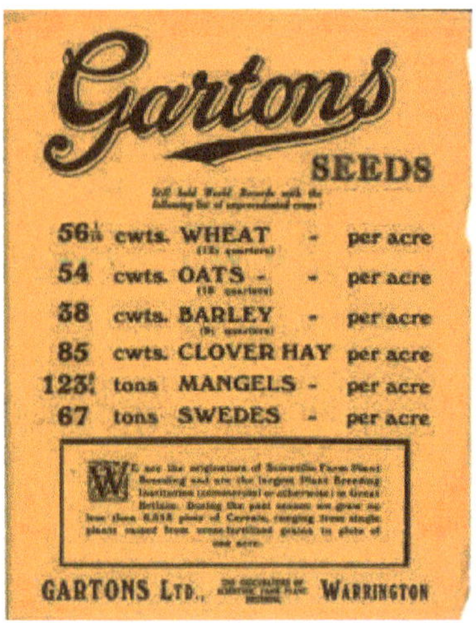

Commercial success and competition

Gartons Limited was the United Kingdom's only major agricultural plant breeding company. But this caused them difficulties as early as in their Spring 1900 seed catalogue where a paragraph of the introduction reads:

It has come to our knowledge that nearly all the New Breeds introduced by us up to the present time have been renamed by various dealers and are being offered by them under different names. Although the honesty of this conduct is more than questionable, we are resigned for the present to regard it as a novel form of flattery, but we strongly recommend all those who wish to secure our Seeds to order them direct from us, as they cannot be procured from any other genuine source.

After the Great War (1914–1918) the United Kingdom government funded cereal breeding at the Plant Breeding Institute at Cambridge which had been founded in 1912, and funded the setting up of plant breeding stations in Edinburgh (1921), Aberystwyth (1919) and in Glasnevin, Northern Ireland in competition with Gartons Limited.

The 1960 Report of the Committee on Transactions in Seeds set up by Parliament entitled Plant Breeders' Rights stated that whilst two thirds of breeding work was by then carried out by government organisations, one third was in the hands of private breeders. And yet the only non-government funded agricultural crop plant breeding, research and testing establishment visited by the Committee was to Gartons Limited. The United Kingdom's Plant Varieties and Seeds Act 1964 allowed plant breeders to fully protect and be rewarded for their introductions. The last variety bred by Gartons, 'Apex' wheat, was the first British bred wheat to be awarded plant breeders rights in 1967 under this legislation.

Chairmen and Directors

The eldest of the three Garton brothers, Robert Garton, was the first chairman of Gartons Limited. He died in February 1950, aged 91, a widower with no children. He was succeeded by the youngest brother, Thomas R. Garton, who died in May 1956. His son, John, was chairman from August 1963 until September 1965. Dr John Garton, the middle of the three brothers, was never, a director.

Nine members of the Miln family were involved with the business over a period of seventy five years.

MR. GEORGE P. MILN, J.P., F.L.S.

George Peddie Miln

Born at Linlathen, Broughty Ferry in 1861, George Peddie Miln trained in a Dundee seed warehouse, the traditional Scottish training for a young man with ambition in the seed trade. He moved to Chester and ran one of its old established seed merchants before joining Gartons Limited as its first Managing Director in 1898. He was a member of Her Majesty's Board of Agriculture Seeds Advisory Council during the first World War. Both the Seeds Act 1920 and the formation of the National Institute of Agricultural Botany came about with his considerable encouragement during his three-year presidency of the Seed Trade Association of the United Kingdom. Of his eleven children, five trained in the seed trade. He was a Justice of the Peace in the City of Chester, a Fellow of the Linnean Society and a member of the Council of the Royal Agricultural Society of England. He died aged 68 following an unsuccessful operation.

When George P. Miln died in 1928 he was succeeded as managing director by his eldest son, Thomas Edward Miln (1890–1963) who for over twenty five years was chairman of the Retail Committee of the Seed Trade Association of the United Kingdom which proudly kept its independence from government control during World War II. A plant breeder as well as a businessman he is credited with the introduction of the sugar beet crop to UK agriculture.

When Gartons Limited became a public quoted company on the London Stock Exchange in 1947 T. E. Miln entered into a further ten year employment contract as Managing Director, even though he was already 59, such was his reputation. Both his sons, Wallace and George, joined him in the business.

T.E. Miln

Wallace Miln

T. E. Miln was succeeded in 1961 by his elder son, Wallace Miln (1919–1994) also a skilled plant breeder and seed analyst.Wallace Miln was one of the three founders of the British Association of Plant Breeders at the time of the introduction of the United Kingdom's Plant Varieties and Seeds Act 1964. He was twice President of the Seed Trade Association of United Kingdom. He left Gartons Limited in 1973 to join his elder son, Barnaby Miln, who had trained at Gartons and later with Northrup-King in Minneapolis before setting up his own seed business at Bodenham in Herefordshire.

From 1947 Gartons Limited's shares were quoted on the London Stock Exchange. The Company's profits for the previous seven years had averaged £48,940. In 1949 the profit of £75,340 was the highest ever recorded by the Company.

Decline and Closure

In 1965 Peter Darlington became chairman of Gartons Limited. In a dramatic change of direction in 1967 Gartons Limited ceased retailing seeds directly to farmers. Instead a new brand was created, Gartons GROplan, and marketing became wholesale through agricultural merchants throughout the United Kingdom. However Gartons GROplan was sold to Agricultural Holdings Company Limited in 1971, but Gartons Limited continued as a plant breeding company.

Gartons plc, as it became known, ceased trading in 1983. The seed warehouse at Bridge Foot, Warrington, which had dominated the town centre skyline since 1910, was demolished in October 1986.A hotel was built on the site.

Varieties Bred and Introduced

Barley Varieties:

Barley varieties bred and introduced to UK agriculture include Standwell in 1898, Invincible (1899), Zero (1900), Brewer's Favourite (1901), The Maltster (1903), Eclipse (1904), Ideal (1906), 1917 (1918), Admiral Beatty (1920), Triumphant (1927).

Oat Varieties:

Oat varieties bred and introduced to UK agriculture include Abundance in 1892, Pioneer (1899), Tartar King (1899), Waverley (1900), Goldfinder (1901), Storm King (1902), Excelsior (1903), Colossal (1904), Rival (1906), Unversed (1907), Bountiful (1908), The Yielder (1909), The Record (1911), The Leader (1913), Supreme (1915), The Hero (1916), The Captain (1919), Sir Douglas Haig (1920), Marvellous (1921), Superb (1923), Earl Haig (1925), Cropwell (1926), Plentiful (1927), Black Prince (1929), Progress (1930), Unique (1931), Onward (1935), Jubilee (1936), Royal Scot (1940), Spitfire (1945), Early Grey (1946), Forward (1953), Angus (1959).

Wheat Varieties:

Wheat varieties bred and introduced to UK agriculture include White Monarch in 1899, White Pearl (1900), Red King (1900), New Era (1903), Reliance (1909), Victor (1910), Benefactor (1914), Early Cone (1918), The Hawk (1918), Marshal Foch (1919), Rector (1923), Benefactress (1925), Renown (1926), Wilhelmina Regenerated (1928), Gartons No 60 (1932), Gartons Q3 (1933), Redman (1934), Little Tich (1935), Wilma (1936), Warden (1938), Meteor (1941), Pilot (1945), Welcome (1950), Masterpiece (1951), Alpha (1952), Victor II (1953), Ritchie (1957), Apex (1965).

Swede Varieties:

Swede varieties bred and introduced to UK agriculture include Zero in 1900, Lord Derby (1900), Perfection (1900), Monarch (1900), Model (1900), Green Tankard (1901),

Keepwell (1902), Cropwell (1903), Superlative (1905), Victory (1907), Incomparable (1907), Warrington (1914), Acme (1914), Magnificent (1917), Viking (1918), Feedwell (1922), White Fleshed (1933), Parkside (1951), Townhead (1951).

Turnip Varieties:

Turnip varieties bred and introduced to UK agriculture include Mammoth Purpletop in 1900, Greentop Scotch Yellow (1900), Hardy Green Globe (1900), Pioneer (1903), Purpletop Long Keeping (1912), Deep Golden Yellow Long Keeping (1912), The Bruce (1917), The Grampian (1920), The Wallace (1935).

Sugar Beet Varieties:

Sugar Beet varieties bred and introduced to UK agriculture include Gartons in 1909, Gartons C (1941) and Gartons Number 632 (1962).

Kale and Kail Varieties:

Kale varieties bred and introduced to UK agriculture include Thousand Headed in 1902, Marrow Stem Kail (1912), Gartons Hybrid (1937) and Hungry Gap (1941).

Kohl Rabi Varieties:

Kohl Rabi varieties bred and introduced to UK agriculture include Large Green in 1902 and Improved Short Top in 1904.

Mangel Varieties:

Mangel varieties bred and introduced to UK agriculture include Large Yellow Intermediate in 1900, Mammoth Long Red (1900), Golden Tankard (1900), Large Yellow Globe (1900), Select Golden Globe (1900), Sugar (1905), Red Intermediate (1905), Devon Yellow Intermediate (1907), Golden Gatepost (1909), Large Red Globe (1910), Large Golden Globe (1910), Nonsuch (1917), Sunrise (1919), White Knight (1922), New Combination (1924), Lemon Globe (1927), Gartons Number 432 (1928), Gartons Number 47 (1931), White Chief (1935), Gartons Number 601 (1960).

Rape Varieties:

Rape varieties bred and introduced to UK agriculture include Broadleaved in 1906, Early Giant (1947) and Late Dwarf (1947).

Herbage Grass Varieties: Herbage grass varieties bred and introduced to UK agriculture include Hatchmere Perennial Ryegrass in 1899, Ellesmere Perennialized Italian Ryegrass (1907), Pickmere Perennial Ryegrass (1932), Delamere Cocksfoot (1936), Oakmere Timothy (1940), Flaxmere (1952), Gartons Tall Fescue (1955), Marbury Meadow Fescue (1957), Barmere Timothy (1958).

Clover Varieties

Clover varieties bred and introduced to UK agriculture include Giant Cowgrass in 1898, Perennial Cowgrass (1898), Perennialized Broad Red Clover (1898), Gartons White Clover (1898) and Broad Red Clover (1907).

Field Cabbage Varieties

Field Cabbage varieties bred and introduced to UK agriculture include Early Ox Heart in 1900, Extra Early Express (1900), Early Drumhead (1900), Selected Drumhead Savoy (1902), Selected Ormskirk Savoy (1902), Gartons Cattle Drumhead (1904), Giant Purple Flat Poll (1917), Utility (1924), Intermediate Drumhead (1924), Gartons Primo (1939).

Field Carrot Varieties

Field Carrot varieties bred and introduced to UK agriculture include Scarlet Intermediate in 1900, Mid Season Scarlet (1911), Mammoth White (1924), Intermediate Stump Rooted (1935), Red Cored Early Market (1935), Short Stump Rooted (1938), Giant White (1939).

Lupin, Parsnip, Potato, Sprouting Broccoli, Winter Beans and Winter Rye Varieties

Other crop varieties bred and introduced to UK agriculture include Gartons Lupin in 1922, Gartons Field Parsnips (1902), Gartons Number 12 Potato (1912), Gartons Purple Sprouting Broccoli (1903), Gartons Giant Winter Bean (1922), GS Giant Winter Bean (1950), P/L 14 Giant Winter Bean (1954), Gartons Giant Large Grained Winter Rye (1922).

NORMAN BORLAUG

Norman Ernest Borlaug (March 25, 1914 – September 12, 2009) was an American biologist and humanitarian who led initiatives worldwide that contributed to the extensive increases in agricultural production termed the Green Revolution. Borlaug has been awarded multiple honors for his work, including the Nobel Peace Prize, the Presidential Medal of Freedom and the Congressional Gold Medal.

Borlaug received his B.Sc. Biology in 1937 and Ph.D. in plant pathology and genetics from the University of Minnesota in 1942. He took up an agricultural research position in Mexico, where he developed semi-dwarf, high-yield, disease-resistant wheat varieties. During the mid-20th century, Borlaug led the introduction of these high-yielding varieties combined with modern agricultural production techniques to Mexico, Pakistan, and India. As a result, Mexico became a net exporter of wheat by 1963. Between 1965 and 1970, wheat yields nearly doubled in Pakistan and India, greatly improving the food security in those nations. Borlaug is often called «the father of the Green Revolution», and is credited with saving over a billion people worldwide from starvation. According to Jan Douglas, executive assistant to the president of the World Food Prize Foundation, the source of this number is Gregg Easterbrook›s 1997 article «Forgotten Benefactor of Humanity», the article states that the «form of agriculture that Borlaug preaches may have prevented a billion deaths.» He was awarded the Nobel Peace Prize in 1970 in recognition of his contributions to world peace through increasing food supply. Later in his life, he helped apply these methods of increasing food production in Asia and Africa.

Early Life, Education and Family

Norman Borlaug wrestling at the University of Minnesota

Borlaug was the great-grandchild of Norwegian immigrants to the United States. Ole Olson Dybevig and Solveig Thomasdatter Rinde, from Feios, a small village in Vik kommune, Sogn og Fjordane, Norway, emigrated to Dane County, Wisconsin in 1854. The family eventually moved to the small Norwegian-American community of Saude, near Cresco, Iowa. There they were members of Saude Lutheran Church, where Norman was both baptized and confirmed.

The eldest of four children — his three younger sisters were Palma Lillian (Behrens; 1916–2004), Charlotte (Culbert; b. 1919) and Helen (1921–1921) — Borlaug was born to Henry Oliver (1889–1971) and Clara (Vaala) Borlaug (1888–1972) on his grandparents' farm in Saude in 1914. From age seven to nineteen, he worked on the 106-acre (43 ha) family farm west of Protivin, Iowa, fishing, hunting, and raising corn, oats, timothy-grass, cattle, pigs and chickens. He attended the one-teacher, one-room New Oregon #8 rural school in Howard County, through eighth grade. Today, the school building, built in 1865, is owned by the Norman Borlaug Heritage Foundation as part of "Project Borlaug Legacy". At Cresco High School, Borlaug played on the football, baseball and wrestling teams, on the latter of which his coach, Dave Barthelma, continually encouraged him to "give 105%". He attributed his decision to leave the farm and pursue further education to his grandfather, Nels Olson Borlaug (1859–1935), who strongly encouraged Borlaug's learning, oncc saying, "You're wiser to fill your head now if you want to fill your belly later on." Through a Depression-era program known as the National Youth Administration, he was able to enroll at the University of Minnesota in 1933. Borlaug failed the entrance exam, but was accepted to the school's newly created two-year General College. After two quarters, he transferred to the College of Agriculture's forestry program. While at the University of Minnesota, he was a member of the varsity wrestling team, reaching the Big Ten semifinals and helped introduce the sport to Minnesota high schools by putting on exhibition matches around the state.

Wrestling taught me some valuable lessons ... I always figured I could hold my own against the best in the world. It made me tough. Many times, I drew on that strength. It's an inappropriate crutch perhaps, but that's the way I'm made.

To finance his studies, Borlaug had to put his education on hold periodically to take a job. One of these jobs, in 1935, was as a leader in the Civilian Conservation Corps, working with the unemployed on U.S. federal projects. Many of the people who worked for him were starving. He later recalled, "I saw how food changed them ... All of this left scars on me". From 1935 to 1938, before and after receiving his bachelor of science in forestry in 1937, Borlaug worked for the United States Forest Service at stations in Massachusetts and Idaho. He spent one summer in the middle fork of Idaho's Salmon River, the most isolated piece of wilderness in the lower 48 states at the time.

In the last months of his undergraduate education, Borlaug attended a Sigma Xi lecture by Elvin Charles Stakman, a professor and soon-to-be head of the plant pathology group at the University of Minnesota. The event was pivotal for Borlaug's future. Stakman, in his speech titled "These Shifty Little Enemies that Destroy our Food Crops", discussed the manifestation of the plant disease rust, a parasitic fungus that feeds on phytonutrients in wheat, oat, and barley crops across the U.S. He had discovered that special plant breeding methods created plants resistant to rust. His research greatly interested Borlaug, and when Borlaug's job at the Forest Service was eliminated because of budget cuts, he asked Stakman if he should go into forest pathology. Stakman advised him to focus on plant pathology instead. Borlaug subsequently enrolled at the University to study plant pathology under Stakman, receiving a master of science degree in 1940 and Ph.D. in plant pathology and genetics in 1942.

Borlaug was a member of Alpha Gamma Rho fraternity. While in college, he met his future wife, Margaret Gibson, as he waited tables at a university Dinkytown coffee shop where they both worked. They had three children, Norma Jean "Jeanie" Laube, Scotty (who died soon after birth from spina bifida), and William Borlaug; five grandchildren, and six great-grandchildren. On March 8, 2007, Margaret Borlaug died at the age of 95, following a fall. They had been married for 69 years. Borlaug spent the last years of his life in northern Dallas although, as a result of his global humanitarian efforts, he actually resided there only a few weeks of the year.

Career

From 1942 to 1944, Borlaug was employed as a microbiologist at DuPont in Wilmington, Delaware. It was planned that he would lead research on industrial and agricultural bacteriocides, fungicides, and preservatives. However, following the December 7, 1941, attack on Pearl Harbor Borlaug tried to enlist in the military, but was rejected under wartime labor regulations; his lab was converted to conduct research for the United States armed forces. One of his first projects was to develop glue that could withstand the warm salt water of the South Pacific. The Imperial Japanese Navy had gained control of the island of Guadalcanal, and patrolled the sky and sea by day. The only way for U.S. forces to supply the troops stranded on the island was to approach at night by speedboat, and jettison boxes of canned food and other supplies into the surf to wash ashore. The problem was that the glue holding these containers together disintegrated in saltwater. Within weeks, Borlaug and his colleagues had developed an adhesive that resisted corrosion, allowing food and supplies to reach the stranded Marines. Other tasks included work with camouflage; canteen disinfectants; DDT to control malaria; and insulation for small electronics.

In 1940, the Avila Camacho administration took office in Mexico. The administration's primary goal for Mexican agriculture was augmenting the nation's industrialization and economic growth. U.S. Vice President-Elect Henry Wallace, who was instrumental in persuading the Rockefeller Foundation to work with the Mexican government in agricultural development, saw Avila Camacho's ambitions as beneficial to U.S. economic and military interests. The Rockefeller Foundation contacted E.C. Stakman and two other leading agronomists. They developed a proposal for a new organization, the Office of Special Studies, as part of the Mexican Government, but directed by the Rockefeller Foundation. It was to be staffed with both Mexican and US scientists, focusing on soil development, maize and wheat production, and plant pathology.

Stakman chose Dr. Jacob George "Dutch" Harrar as project leader. Harrar immediately set out to hire Borlaug as head of the newly established Cooperative Wheat Research and Production Program in Mexico; Borlaug declined, choosing to finish his war service at DuPont. In July 1944, after rejecting DuPont's offer to double his salary, and temporarily leaving behind his pregnant wife and 14-month-old daughter, he flew to Mexico City to head the new program as a geneticist and plant pathologist. In 1964, he was made the director of the International Wheat Improvement Program at El Batán, Texcoco, on the eastern fringes of Mexico City, as part of the newly established Consultative Group on International Agricultural Research's International Maize and Wheat Improvement Center*(Centro Internacional de Mejoramiento de Maíz y Trigo,* or CIMMYT). Funding for this autonomous international research training institute developed from the Cooperative Wheat Research Production Program was undertaken jointly by the Ford and Rockefeller Foundations and the Mexican government.

Borlaug retired officially from the position in 1979, but remained a CIMMYT senior consultant. In addition to taking up charitable and educational roles, he continued to be involved in plant research at CIMMYT with wheat, triticale, barley, maize, and high-altitude sorghum.

In 1981, Borlaug became a founding member of the World Cultural Council.

In 1984, Borlaug began teaching and conducting research at Texas A&M University. Eventually being given the title Distinguished Professor of International Agriculture at the university and the holder of the Eugene Butler Endowed Chair in Agricultural Biotechnology, Borlaug remained at A&M until his death in September 2009.

Wheat Research in Mexico

The Cooperative Wheat Research Production Program, a joint venture by the Rockefeller Foundation and the Mexican Ministry of Agriculture, involved research in genetics, plant breeding, plant pathology, entomology, agronomy, soil science, and cereal technology. The goal of the project was to boost wheat production in Mexico, which at the time was importing a large portion of its grain. Plant pathologist George Harrar recruited and assembled the wheat research team in late 1944. The four other members were soil scientist William Colwell; maize breeder Edward Wellhausen; potato breeder John Niederhauser; and Norman Borlaug, all from the United States. During the sixteen years Borlaug remained with the project, he bred a series of remarkably successful high-yield, disease-resistant, semi-dwarf wheat.

Borlaug said that his first few years in Mexico were difficult. He lacked trained scientists and equipment. Local farmers were hostile towards the wheat program because of serious crop losses from 1939 to 1941 due to stem rust. "It often appeared to me that I had made a dreadful mistake in accepting the position in Mexico," he wrote in the epilogue to his book, *Norman Borlaug on World Hunger*. He spent the first ten years breeding wheat cultivars resistant to disease, including rust. In that time, his group made 6,000 individual crossings of wheat.

Wheat is the third most-produced cereal crop.

Double Wheat Season

Initially, Borlaug's work had been concentrated in the central highlands, in the village of Chapingo near Texcoco, where the problems with rust and poor soil were most prevalent. He realized that he could speed up breeding by taking advantage of the country›s two growing seasons. In the summer he would breed wheat in the central highlands as usual, then immediately take the seeds north to the Yaqui Valley research station near Ciudad Obregón, Sonora. The difference in altitudes and temperatures would allow more crops to be grown each year.

Borlaug›s boss, George Harrar, was against this expansion. Besides the extra costs of doubling the work, Borlaug›s plan went against a then-held principle of agronomy that has since been disproved. It was believed that to store energy for germination before being planted, seeds needed a rest period after harvesting. When Harrar vetoed his plan, Borlaug resigned. Elvin Stakman, who was visiting the project, calmed the situation, talking Borlaug into withdrawing his resignation and Harrar into allowing the double wheat season. As of 1945, wheat would then be bred at locations 700 miles (1000 km) apart, 10 degrees apart in latitude, and 8500 feet (2600 m) apart in altitude. This was called «shuttle breeding».

Locations of Borlaug's research stations in the Yaqui Valley and Chapingo.

As an unexpected benefit of the double wheat season, the new breeds did not have problems with photoperiodism. Normally, wheat varieties cannot adapt to new environments, due to the changing periods of sunlight. Borlaug later recalled, "As it worked out, in the north, we were planting when the days were getting shorter, at low elevation and high temperature. Then we'd take the seed from the best plants south and plant it at high elevation, when days were getting longer and there was lots of rain. Soon we had varieties that fit the whole range of conditions. That wasn't supposed to happen by the books". This meant that the project would not need to start separate breeding programs for each geographic region of the planet.

Increasing Disease Resistance through Multiline Varieties

Because pure line (genotypically identical) plant varieties often only have one or a few major genes for disease resistance, and plant diseases such as rust are continuously producing new races that can overcome a pure line's resistance, multiline varieties were developed. Multiline varieties are mixtures of several phenotypically similar pure lines which each have different genes for disease resistance. By having similar heights, flowering and maturity dates, seed colors, and agronomic characteristics, they remain compatible with each other, and do not reduce yields when grown together on the field.

In 1953, Borlaug extended this technique by suggesting that several pure lines with different resistance genes should be developed through backcross methods using one recurrent parent. Backcrossing involves crossing a hybrid and subsequent generations with a recurrent parent. As a result, the genotype of the backcrossed progeny becomes increasingly similar to that of the recurrent parent. Borlaug›s method would allow the various different disease-resistant genes from several donor parents to be transferred into a single recurrent parent. To make sure each line has different resistant genes, each donor parent is used in a separate backcross program. Between five and ten of these lines may then be mixed depending upon the races of pathogen present in the region. As this process is repeated, some lines will become susceptible to the pathogen. These lines can easily be replaced with new resistant lines. As new sources of resistance become available, new lines are developed. In this way, the loss of crops is kept to a minimum, because only one or a few lines become susceptible to a pathogen within a given season, and all other crops are unaffected by the disease. Because the disease would spread more slowly than if the entire population were susceptible, this also reduces the damage to susceptible lines. There is still the possibility that a new race of pathogen will develop to which all lines are susceptible, however.

Dwarfing

Dwarfing is an important agronomic quality for wheat; dwarf plants produce thick stems. The cultivars Borlaug worked with had tall, thin stalks. Taller wheat grasses better compete for sunlight, but tend to collapse under the weight of the extra grain — a trait called lodging — from the rapid growth spurts induced by nitrogen fertilizer Borlaug used in the poor soil. To prevent this, he bred wheat to favor shorter, stronger stalks that could better support larger seed heads. In 1953, he acquired a Japanese dwarf variety of wheat called Norin 10 developed by Orville Vogel, that had been crossed with a high-yielding American cultivar called Brevor 14. Norin 10/Brevor 14 is semi-dwarf (one-half to two-thirds the height of standard varieties) and produces more stalks and thus more heads of grain per plant. Also, larger amounts of assimilate were partitioned into the actual grains, further increasing the yield. Borlaug crossbred the semi-dwarf Norin 10/Brevor 14 cultivar with his disease-resistant cultivars to produce wheat varieties that were adapted to tropical and sub-tropical climates.

Borlaug's new semi-dwarf, disease-resistant varieties, called Pitic 62 and Penjamo 62, changed the potential yield of spring wheat dramatically. By 1963, 95% of Mexico's wheat crops used the semi-dwarf varieties developed by Borlaug. That year, the harvest was six times larger than in 1944, the year Borlaug arrived in Mexico. Mexico had become fully self-sufficient in wheat production, and a net exporter of wheat. Four other high-yield varieties were also released, in 1964: Lerma Rojo 64, Siete Cerros, Sonora 64, and Super X.

Expansion to South Asia: the Green Revolution

In 1961 to 1962, Borlaug's dwarf springs wheat strains were sent for multilocation testing in the International Wheat Rust Nursery, organized by the U.S. Department of Agriculture. In March 1962, a few of these strains were grown in the fields of the Indian Agricultural Research Institute in Pusa, New Delhi, India. In May 1962, M. S. Swaminathan, a member of IARI's wheat program, requested of Dr. B. P. Pal, director of IARI, to arrange for the visit of Borlaug to India and to obtain a wide range of dwarf wheat seed possessing the Norin 10 dwarfing genes. The letter was forwarded to the Indian Ministry of Agriculture headed by Shri C. Subramaniam, which arranged with the Rockefeller Foundation for Borlaug's visit. In March 1963, the Rockefeller Foundation and the Mexican government sent Borlaug and Dr. Robert Glenn Anderson to India to continue his work. He supplied 100 kg (220 lb) of seed from each of the four most promising strains and 630 promising selections in advanced generations to the IARI in October 1963, and test plots were subsequently planted at Delhi, Ludhiana, Pant Nagar, Kanpur, Pune and Indore. Anderson stayed as head of the Rockefeller Foundation Wheat Program in New Delhi until 1975.

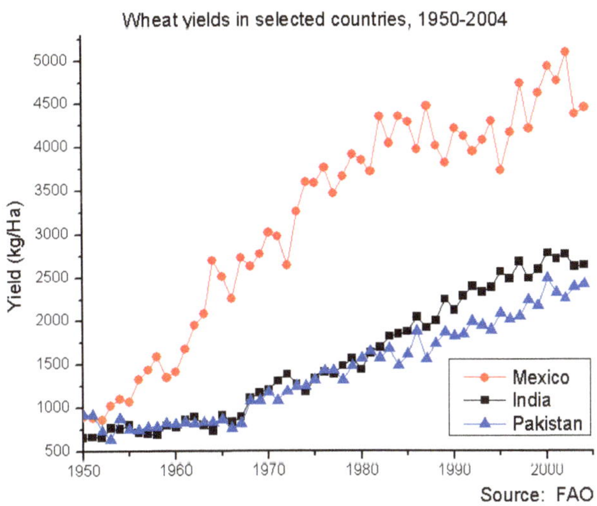

Wheat yields in Mexico, India and Pakistan, 1950 to 2004. Baseline is 500 kg/ha.

During the mid-1960s, the Indian subcontinent was at war and experienced minor famine and starvation, which was limited partially by the U.S. shipping a fifth of its wheat production to India in 1966 & 1967. The Indian and Pakistani bureaucracies and the region's cultural opposition to new agricultural techniques initially prevented Borlaug from fulfilling his desire to immediately plant the new wheat strains there. In 1965, as a response to food shortages, Borlaug imported 550 tons of seeds for the government.

Biologist Paul R. Ehrlich wrote in his 1968 bestseller *The Population Bomb*, "The battle to feed all of humanity is over ... In the 1970s and 1980s hundreds of millions of people will starve to death in spite of any crash programs embarked upon now." Ehrlich said, "I have yet to meet anyone familiar with the situation who thinks India will be self-sufficient in food by 1971," and "India couldn't possibly feed two hundred million more people by 1980."

In 1965, after extensive testing, Borlaug's team, under Anderson, began its effort by importing about 450 tons of Lerma Rojo and Sonora 64 semi-dwarf seed varieties: 250 tons went to Pakistan and 200 to India. They encountered many obstacles. Their first shipment of wheat was held up in Mexican customs and so it could not be shipped from the port at Guaymas in time for proper planting. Instead, it was sent via a 30-truck convoy from Mexico to the U.S. port in Los Angeles, encountering delays at the Mexico - United States border. Once the convoy entered the U.S., it had to take a detour, as the U.S. National Guard had closed the freeway due to Watts riots in Los Angeles. When the seeds reached Los Angeles, a Mexican bank refused to honor Pakistan treasury›s payment of US$100,000, because the check contained three misspelled words. Still, the seed was loaded onto a freighter destined for Bombay, India, and Karachi, Pakistan. Twelve hours into the freighter›s voyage, war broke out between India and Pakistan over the Kashmir region. Borlaug received a telegram from the Pakistani minister of agriculture, Malik Khuda Bakhsh Bucha: «I›m sorry to hear you are having trouble with my check, but I›ve got troubles, too. Bombs are falling on my front lawn. Be patient, the money is in the bank ...»

These delays prevented Borlaug›s group from conducting the germination tests needed to determine seed quality and proper seeding levels. They started planting immediately, and often worked in sight of artillery flashes. A week later, Borlaug discovered that his seeds were germinating at less than half the normal rate. It later turned out that the seeds had been damaged in a Mexican warehouse by over-fumigation with a pesticide. He immediately ordered all locations to double their seeding rates.

The initial yields of Borlaug›s crops were higher than any ever harvested in South Asia. The countries subsequently committed to importing large quantities of both the Lerma Rojo 64 and Sonora 64 varieties. In 1966, India imported 18,000 tons —the largest purchase and import of any seed in the world at that time. In 1967, Pakistan imported 42,000 tons, and Turkey 21,000 tons. Pakistan›s import, planted on 1.5 million acres (6,100 km^2), produced enough wheat to seed the entire nation›s wheatland the following year. By 1968, when Ehrlich›s book was released, William Gaud of the United States Agency for International Development was calling Borlaug›s work a «Green Revolution». High yields led to a shortage of various utilities — labor to harvest the crops, bullock carts to haul it to the threshing floor, jute bags, trucks, rail cars, and grain storage facilities. Some local governments were forced to close school buildings temporarily to use them for grain storage.

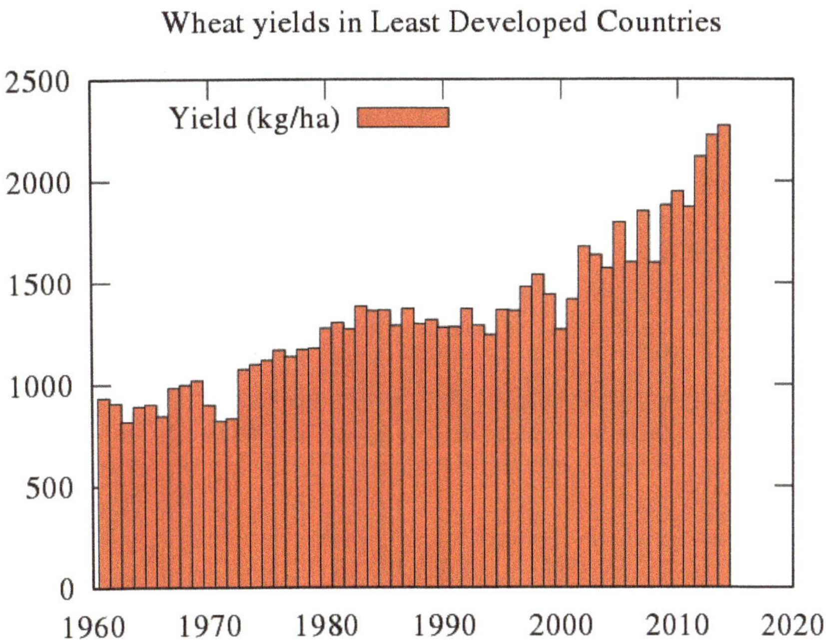

Wheat yields in least developed countries since 1961.

In Pakistan, wheat yields nearly doubled, from 4.6 million tons in 1965 to 7.3 million tons in 1970; Pakistan was self-sufficient in wheat production by 1968. Yields were over 21 million tons by 2000. In India, yields increased from 12.3 million tons in 1965 to 20.1 million tons in 1970. By 1974, India was self-sufficient in the production of all cereals. By 2000, India was harvesting a record 76.4 million tons (2.81 billion bushels) of wheat. Since the 1960s, food production in both nations has increased faster than the rate of population growth. India's use of high-yield farming has prevented an estimated 100 million acres (400,000 km²) of virgin land from being converted into farmland—an area about the size of California, or 13.6% of the total area of India. The use of these wheat varieties has also had a substantial effect on production in six Latin American countries, six countries in the Near and Middle East, and several others in Africa.

Borlaug›s work with wheat contributed to the development of high-yield semi-dwarf *indica* and *japonica* rice cultivars at the International Rice Research Institute and China's Hunan Rice Research Institute. Borlaug's colleagues at the Consultative Group on International Agricultural Research also developed and introduced a high-yield variety of rice throughout most of Asia. Land devoted to the semi-dwarf wheat and rice varieties in Asia expanded from 200 acres (0.8 km²) in 1965 to over 40 million acres (160,000 km²) in 1970. In 1970, this land accounted for over 10% of the more productive cereal land in Asia.

Nobel Peace Prize

For his contributions to the world food supply, Borlaug was awarded the Nobel Peace Prize in 1970. Norwegian officials notified his wife in Mexico City at 4:00 am, but Borlaug had already left for the test fields in the Toluca valley, about 40 miles (65 km) west of Mexico City. A chauffeur took her to the fields to inform her husband. According to his daughter, Jeanie Laube, "My mom said, 'You won the Nobel Peace Prize,' and he said, 'No, I haven't', It took some convincing ... He thought the whole thing was a hoax". He was awarded the prize on December 10. In his Nobel Lecture the following day, he speculated on his award: "When the Nobel Peace Prize Committee designated me the recipient of the 1970 award for my contribution to the 'green revolution', they were in effect, I believe, selecting an individual to symbolize the vital role of agriculture and food production in a world that is hungry, both for bread and for peace". His speech repeatedly presented improvements in food production within a sober understanding of the context of population. "The green revolution has won a temporary success in man's war against hunger and deprivation; it has given man a breathing space. If fully implemented, the revolution can provide sufficient food for sustenance during the next three decades. But the frightening power of human reproduction must also be curbed; otherwise the success of the green revolution will be ephemeral only. Most people still fail to comprehend the magnitude and menace of the "Population Monster"...Since man is potentially a rational being, however, I am confident that within the next two decades he will recognize the self-destructive course he steers along the road of irresponsible population growth..."

Borlaug Hypothesis

Borlaug continually advocated increasing crop yields as a means to curb deforestation. The large role he played in both increasing crop yields and promoting this view has led to this methodology being called by agricultural economists the "Borlaug hypothesis", namely that *increasing the productivity of agriculture on the best farmland can help control deforestation by reducing the demand for new farmland*. According to this view, assuming that global food demand is on the rise, restricting crop usage to traditional low-yield methods would also require at least one of the following: the world population to decrease, either voluntarily or as a result of mass starvations; or the conversion of forest land into crop land. It is thus argued that high-yield techniques are ultimately saving ecosystems from destruction. On a global scale, this view holds strictly true *ceteris paribus*, if deforestation only occurs to increase land for agriculture. But other land uses exist, such as urban areas, pasture, or fallow, so further research is necessary to ascertain what land has been converted for what purposes, to determine how true this view remains. Increased profits from high-yield production may also induce cropland expansion in any case, although as world food needs decrease, this expansion may decrease as well.

Criticisms and His View of Critics

Borlaug's name is nearly synonymous with the Green Revolution, against which many criticisms have been mounted over the decades by environmentalists and some nutritionalists.

Throughout his years of research, Borlaug's programs often faced opposition by people who consider genetic crossbreeding to be unnatural or to have negative effects. Borlaug's work has been criticized for bringing large-scale monoculture, input-intensive farming techniques to countries that had previously relied on subsistence farming. These farming techniques often reap large profits for U.S. agribusiness and agrochemical corporations and have been criticized for widening social inequality in the countries owing to uneven food distribution while forcing a capitalist agenda of U.S. corporations onto countries that had undergone land reform.

Other concerns of his critics and critics of biotechnology in general include: that the construction of roads in populated third-world areas could lead to the destruction of wilderness; the crossing of genetic barriers; the inability of crops to fulfill all nutritional requirements; the decreased biodiversity from planting a small number of varieties; the environmental and economic effects of inorganic fertilizer and pesticides; the amount of herbicide sprayed on fields of herbicide-resistant crops.

Borlaug dismissed most claims of critics, but did take certain concerns seriously. He stated that his work has been "a change in the right direction, but it has not transformed the world into a Utopia". Of environmental lobbyists he stated, "some of the environmental lobbyists of the Western nations are the salt of the earth, but many of them are elitists. They've never experienced the physical sensation of hunger. They do their lobbying from comfortable office suites in Washington or Brussels. If they lived just one month amid the misery of the developing world, as I have for fifty years, they'd be crying out for tractors and fertilizer and irrigation canals and be outraged that fashionable elitists back home were trying to deny them these things".

Later Roles

Following his retirement, Borlaug continued to participate in teaching, research and activism. He spent much of the year based at CIMMYT in Mexico, conducting research, and four months of the year serving at Texas A&M University, where he had been a distinguished professor of international agriculture since 1984. From 1994 to 2003, Borlaug served on the International Fertilizer Development Center board of directors. In 1999, the university's Board of Regents named its US$16 million Center for Southern Crop Improvement in honor of Borlaug. He worked in the building's Heep Center, and taught one semester each year.

Production in Africa

In the early 1980s, environmental groups that were opposed to Borlaug's methods campaigned against his planned expansion of efforts into Africa. They prompted the Rockefeller and Ford Foundations and the World Bank to stop funding most of his African agriculture projects. Western European governments were persuaded to stop supplying fertilizer to Africa. According to David Seckler, former Director General of the International Water Management Institute, "the environmental community in the 1980s went crazy pressuring the donor countries and the big foundations not to support ideas like inorganic fertilizers for Africa."

In 1984, during the Ethiopian famine, Ryoichi Sasakawa, the chairman of the Japan Shipbuilding Industry Foundation (now the Nippon Foundation), contacted the semi-retired Borlaug, wondering why the methods used in Asia were not extended to Africa, and hoping Borlaug could help. He managed to convince Borlaug to help with this new effort, and subsequently founded the Sasakawa Africa Association (SAA) to coordinate the project.

Nigerian exchange students meet Norman Borlaug (third from right) at the World Food seminar, 2003.

The SAA is a research and extension organization that aims to increase food production in African countries that are struggling with food shortages. "I assumed we'd do a few years of research first," Borlaug later recalled, "but after I saw the terrible circumstances there, I said, 'Let's just start growing'." Soon, Borlaug and the SAA had projects in seven countries. Yields of maize in developed African countries tripled. Yields of wheat, sorghum, cassava, and cowpeas also increased in these countries. At present, program activities are under way in Benin, Burkina Faso, Ethiopia, Ghana, Guinea, Mali, Malawi, Mozambique, Nigeria, Tanzania, and Uganda.

From 1986 to 2009, Borlaug was the President of the SAA. That year, a joint venture between The Carter Center and SAA was launched called Sasakawa-Global 2000 (SG 2000). The program focuses on food, population and agricultural policy. Since then, more than 8 million small-scale farmers in 15 African countries have been trained in SAA farming techniques, which have helped them to double or triple grain production. Those elements that allowed Borlaug's projects to succeed in India and Pakistan, such as well-organized economies and transportation and irrigation systems, are severely lacking throughout Africa, posing additional obstacles to increasing yields. Because of this, Borlaug's initial projects were restricted to developed regions of the continent.

Despite these setbacks, Borlaug found encouragement. Visiting Ethiopia in 1994, Jimmy Carter won Prime Minister Meles Zenawi's support for a campaign seeking to aid farmers, using the fertilizer diammonium phosphate and Borlaug's methods. The following season, Ethiopia recorded the largest harvests of major crops in history, with a 32% increase in production, and a 15% increase in average yield over the previous season. For Borlaug, the rapid increase in yields suggested that there was still hope for higher food production throughout sub-Saharan Africa, despite lingering questions and the absence of long-term studies.

World Food Prize

The World Food Prize is an international award recognizing the achievements of individuals who have advanced human development by improving the quality, quantity or availability of food in the world. The prize was created in 1986 by Norman Borlaug, as a way to recognize personal accomplishments, and as a means of education by using the Prize to establish role models for others. The first prize was given to Borlaug's former colleague, M. S. Swaminathan, in 1987, for his work in India. The next year, Swaminathan used the US$250,000 prize to start the MS Swaminathan Research Foundation for research on sustainable development.

Online Education

At the DuPont Agriculture & Nutrition Media Day held in Des Moines, Iowa, on September 25, 2000, Borlaug announced the launch of Norman Borlaug University, an Internet-based learning company for agriculture and food industry personnel. The University was unable to expand the necessary content or customer base, and since late 2001 has been defunct.

Future of Global Farming and Food Supply

The limited potential for land expansion for cultivation worried Borlaug, who, in March 2005, stated that, "we will have to double the world food supply by 2050." With 85% of future growth in food production having to come from lands already in use, he recommends a multidisciplinary research focus to further increase yields, mainly through increased crop immunity to large-scale diseases, such as the rust fungus, which affects all cereals but rice. His dream was to "transfer rice immunity to cereals such as wheat, maize, sorghum and barley, and transfer bread-wheat proteins (gliadin and glutenin) to other cereals, especially rice and maize".

Borlaug believed that genetically modified organisms (GMO) was the only way to increase food production as the world runs out of unused arable land. GMOs were not inherently dangerous "because we've been genetically modifying plants and animals for a long time. Long before we called it science, people were selecting the best breeds." In a review of Borlaug's 2000 publication entitled *Ending world hunger: the promise of biotechnology and the threat of antiscience zealotry*, the authors argued that Borlaug's warnings were still true in 2010,

GM crops are as natural and safe as today's bread wheat, opined Dr. Borlaug, who also reminded agricultural scientists of their moral obligation to stand up to the antiscience crowd and warn policy makers that global food insecurity will not disappear without this new technology and ignoring this reality global food insecurity would make future solutions all the more difficult to achieve.

—Rozwadowski and Kagale

According to Borlaug, “Africa, the former Soviet republics, and the cerrado are the last frontiers. After they are in use, the world will have no additional sizable blocks of arable land left to put into production, unless you are willing to level whole forests, which you should not do. So, future food-production increases will have to come from higher yields. And though I have no doubt yields will keep going up, whether they can go up enough to feed the population monster is another matter. Unless progress with agricultural yields remains very strong, the next century will experience sheer human misery that, on a numerical scale, will exceed the worst of everything that has come before”.

Besides increasing the worldwide food supply, early in his career Borlaug stated that taking steps to decrease the rate of population growth will also be necessary to prevent food shortages. In his Nobel Lecture of 1970, Borlaug stated, «Most people still fail to comprehend the magnitude and menace of the ‹Population Monster› ... If it continues to increase at the estimated present rate of two percent a year, the world population will reach 6.5 billion by the year 2000. Currently, with each second, or tick of the clock, about 2.2 additional people are added to the world population. The rhythm of increase will accelerate to 2.7, 3.3, and 4.0 for each tick of the clock by 1980, 1990, and 2000, respectively, unless man becomes more realistic and preoccupied about this impending doom. The tick-tock of the clock will continually grow louder and more menacing each decade. Where will it all end?» However, some observers have suggested that by the 1990s Borlaug had changed his position on population control. They point to a quote from the year 2000 in which he stated: «I now say that the world has the technology — either available or well advanced in the research pipeline — to feed on a sustainable basis a population of 10 billion people. The more pertinent question today is whether farmers and ranchers will be permitted to use this new technology? While the affluent nations can certainly afford to adopt ultra-low-risk positions, and pay more for food produced by the so-called ‹organic› methods, the one billion chronically undernourished people of the low income, food-deficit nations cannot.» However, Borlaug remained on the advisory board of Population Media Center, an organization working to stabilize world population, until his death.

REFERENCES

1. Breeding Field Crops. 1995. Sleper and Poehlman. Page 3
2. Piperno, D. R.; Ranere, A. J.; Holst, I.; Iriarte, J.; Dickau, R. (2009). “Starch grain and phytolith evidence for early ninth millennium B.P. maize from the Central Balsas River Valley, Mexico”. PNAS. 106 (13): 5019–5024. doi:10.1073/pnas.0812525106. PMC 2664021 . PMID 19307570.
3. Deppe, Carol (2000). Breed Your Own Vegetable Varieties. Chelsea Green Publishing. |page=237-244
4. “Plant breeding”.
5. Spring Seed Catalogue 1899, Gartons Limited
6. Noel Kingsbury (2009). Hybrid: The History and Science of Plant Breeding. University of Chicago Press. p. 140.

7. Suzie Key; Julian K-C Ma & Pascal MW Drake (1 June 2008). "Genetically modified plants and human health". Journal of the Royal Society of Medecine. pp. 290–298. Retrieved 11 March 2015.

8. Davis, D.R.; Epp, M.D.; Riordan, H.D. (2004). "Changes in USDA Food Composition Data for 43 Garden Crops, 1950 to 1999". Journal of the American College of Nutrition. 23 (6): 669–682. doi:10.1080/07315724.2004.10719409.

9. Costa-Font, J.; Mossialos, E. (2007). "Are perceptions of 'risks' and 'benefits' of genetically modified food (in)dependent?". Food Quality and Preference. 18: 173–182. doi:10.1016/j.foodqual.2005.09.013.

10. Connoly, Kate (2009-04-14). "Germany deals blow to GM crops". The Guardian. Retrieved 2009-06-25.

11. Murphy, Kevin M.; K.G. Campbell; S.R. Lyon; S.S. Jones (2007). "Evidence of varietal adaptation to organic farming systems". Field Crops Research. 102 (3): 172–177. doi:10.1016/j.fcr.2007.03.011.

12. Lammerts van Bueren, E.T.; S.S. Jones; L. Tamm; K.M. Murphy; J.R. Myers; C. Leifert; M.M. Messmer (2010). "The need to breed crop varieties suitable for organic farming, using wheat, tomato and broccoli as examples: A review". NJAS- Wageningen Journal of Life Sciences. 58 (3–4): 193–205. doi:10.1016/j.njas.2010.04.001.

13. Lammerts van Bueren, E. T.; G. Backes; H. de Vriend; H. Ostergard (2010). "The role of molecular markers and marker assisted selection in breeding for organic agriculture". Euphytica. 175: 51–64. doi:10.1007/s10681-010-0169-0.

14. Oldeman, l (1994). "The global extent of soil degradation" (PDF). Soil resilience and sustainable land use. 32 (5967): 818–822. Retrieved 7-11-2013.Check date values in: |access-date= (help)

15. Tester, Mark; Langridge, Peter (February 2010). "Breeding technologies to increase crop production in a changing world". Science. AAAS: American Association for the Advancement of Science. 327 (5967): 818–822. doi:10.1126/science.1183700.

16. Haddad, Lawrence; Godfray, H. Charles J.; Beddington, John R.; Crute, Ian R.; Lawrence, David; Muir, James F.; Pretty, Jules; Robinson, Sherman; Thomas, Sandy M. and Toulmin, Camilla (12 February 2010). "Food security: the challenge of feeding 9 billion people". Science. AAAS: American Association for the Advancement of Science. 327 (5967): 812–818. doi:10.1126/science.1185383. PMID 20110467. Cite uses deprecated parameter |coauthor= (help)

17. Bänziger (2000). "Breeding for drought and nitrogen stress tolerance in maize: from theory to practice". From Theory to Practice: 7–9. Retrieved 7-11-2013. Check date values in: |access-date= (help)

18. Casler, Vogal, M,K (1998-03-31). "Accomplishments and impact from breeding for increased forage nutritional value". Crop Science. 39 (1): 12–20. doi:10.2135/cropsci1999.0011183x003900010003x. Retrieved 7-10. Check date values in: |access-date=, |year= / |date= mismatch (help)

19. PRGA 2008. Participatory Plant Breeding (PPB)

20. Sperling et al. 2001. A Framework For Analizing Participatory Plant Breeding Approaches And Results.
21. Ceccarelli 2001. Decentralized-Participatory Plant Breeding: Adapting Crops to Environments and Clients
22. Thro A & Spillane C (2000). "Biotechnology-assisted Participatory Plant Breeding: Complement or Contradiction?" (PDF). CGIAR Systemwide Program on Participatory Research and Gender Analysis for Technology Development and Institutional Innovation. Working Document No. 4 April 2000: 140.
23. Obituary, Warrington Examiner, 27 May 1922
24. The Nurseryman and Seedsman, 4 January 1919
25. Obituary, Warrington Examiner, 11 February 1950
26. Warrington Examiner, 11 February 1950
27. Royal Highland and Agricultural Society of Scotland Archived 29 August 2013 at the Wayback Machine.
28. Page 22, 1900 Seed Catalogue, Gartons Limited
29. Warrington Guardian front page, 6 August 1898
30. Warrington Examiner, 26 June 1948
31. "Obituary: Emeritus Professor Robert Wallace, An authority on agricultural matters — 37 years in Edinburgh chair". The Glasgow Herald. 18 January 1939.
32. Britain's First Chair of Agriculture, Fleming & Robertson, ISBN 0-902164-33-3 pages 47 and 48
33. page 140 Kingsbury, Noel. Hybrid : the history and science of plant breeding. University of Chicago Press, ISBN 978-0-226-43704-0
34. Page 5, 1902 Catalogue, Gartons Limited
35. Introduction to Gartons Limited's Spring 1899 Seed Catalogue
36. Spring Seed Catalogue 1899, Gartons Limited
37. Gartons
38. Seed Catalogue 1898, Gartons
39. Seed Catalogue 1899, Gartons Limited
40. Seed Catalogue 1899, Gartons Limited
41. Seed Catalogue 1898, Gartons
42. Seed Catalogue 1907, Gartons Limited
43. Seed Catalogue 1967, Gartons Limited
44. University of Edinburgh Senatus Academicus Printed Minutes Volume III Oct. 1920-July 1924 p.295
45. Warrington Guardian 26 August 1922
46. British Medical Journal 29 July 1922 page 191
47. University of Edinburgh Senatus Academicus Printed Minutes Volume III Oct. 1920-July 1924 p.332

48. Gartons Limited Seed Catalogue 1901
49. Gartons Limited Seed Catalogue 1910
50. Photograph in T.E.Miln's Newspaper Cutting Book, Miln Family Cartulary
51. Her Majesty's Stationery Office Report, July 1960
52. Her Majesty's Stationery Office Report, July 1960, page 104
53. Photographs of the Gartons Warrington Warehouse and Office built in 1910, The Warrington Examiner 4 April 1931.
54. Gartons Headquarters Plant Breeding Grounds in Cheshire, from The Warrington Examiner 4 April 1931.
55. Warrington Examiner, 11 February 1950
56. The Field, 25 November 1922
57. Warrington Guardian, 5 January 1962
58. London Corn Circular, 1 August 1947
59. Daily Telegraph, 20 September 1949
60. Financial Times, 16 September 1965
61. The Times newspaper, London, Friday 2 July 1971
62. Financial Times, 19 August 1983
63. Warrington Guardian, 15 October 1986
64. Extracted from Gartons Seed Catalogues
65. Swaminathan, M. S. (2009). "Obituary: Norman E. Borlaug (1914–2009) Plant scientist who transformed global food production". Nature. 461 (7266): 894–894. doi:10.1038/461894a. ISSN 0028-0836.
66. Phillips, R. L. (2013). "Norman Ernest Borlaug. 25 March 1914 -- 12 September 2009". Biographical Memoirs of Fellows of the Royal Society. 59: 59–72. doi:10.1098/rsbm.2013.0012.
67. "Nobel Prize winner Norman Borlaug dies at 95". Associated Press. 13 September 2009. Archived from the original on 17 September 2012. Retrieved 17 September 2012.
68. "Borlaug, father of 'Green Revolution', dead", DAWN.com. 14 September 2009. Retrieved 27 May 2015.
69. Scott Kilman and Roger Thurow. "Father of 'Green Revolution' Dies". The Wall Street Journal. Retrieved 5 June 2013.
70. Dowswell, C. (15 October 2009). "Norman Ernest Borlaug (1914-2009)". Science. 326 (5951): 381. doi:10.1126/science.1182211. Retrieved 5 June 2013.
71. "Norman Borlaug". sciencheroes.com. Retrieved 5 June 2013.
72. MacAray, David (15 October 2013). "The Man Who Saved a Billion Lives". The Huffington Post.
73. The phrase "over a billion lives saved" is often cited by others in reference to Norman Borlaug's work.

74. "Hearings". Agriculture.senate.gov. Archived from the original on 2011-11-06.
75. Easterbrook, Gregg (January 1997). "Forgotten benefactor of humanity". The Atlantic. Retrieved October 25, 2016.
76. Enriquez, Juan (September 2007). "Why Can't We Grow New Energy?". TED. Retrieved 18 September 2012.
77. State Historical Society of Iowa. 2002. FY03 HRDP/REAP Grant Application Approval
78. "Father of the Green Revolution — He Helped Feed the World!".
79. "Iowa Rep. Tom Latham Pays Tribute to Dr. Borlaug". The World Food Prize. 20 March 2008. Archived from the original on 3 July 2008. Retrieved 18 September 2012.[dead link]
80. University of Minnesota. 2005."Borlaug and the University of Minnesota". Archived from the original on 2005-03-10. Retrieved 2005-06-18.
81. "Green Giant". Stuertz, Mark. Dallas Observer. 5 December 2002.
82. "Norman Borlaug's Wife Dies at 95". Associated Press. 8 March 2007. Archived from the original on 18 September 2012. Retrieved 18 September 2012. An assistant to the family says she fell recently and never recovered.
83. Wright, Angus 2005. The Death of Ramón González.
84. Davidson, M.G. 1997. An Abundant Harvest: Interview with Norman Borlaug, Recipient, Nobel Peace Prize, 1970, Common Ground, August 12
85. "About Us". World Cultural Council. Retrieved November 8, 2016.
86. Brown, L. R. 1970. Nobel Peace Prize: developer of high-yield wheat receives award (Norman Ernest Borlaug). Science, 30 October 1970;170(957):518-9.
87. ^ Jump up to:a b University of Minnesota. 2005."Borlaug's Work in Mexico". Archived from the original on 2004-12-26. Retrieved 2005-06-19.
88. Miller, Henry I. (January 2012). "Norman Borlaug: The Genius Behind The Green Revolution". Forbes.
89. Borlaug, N.E. (1953). "New approach to the breeding of wheat varieties resistant to Puccinia graminis tritici". Phytopathology. 43: 467.
90. "AGB 301: Principles and Methods of Plant Breeding". Tamil Nadu Agricultural University.
91. Retiz, L.P. (1970). "New wheats and social progress". Science. 169: 952–955. doi:10.1126/science.169.3949.952.
92. Hedden, P (2003). "The genes of the Green Revolution". Trends in Genetics. 19 (1): 5–9. doi:10.1016/s0168-9525(02)00009-4. PMID 12493241.
93. University of Minnesota. 2005."The Beginning of the Green Revolution". Archived from the original on 2004-12-27. Retrieved 2005-06-22.
94. Erlich, Paul: The Population Bomb. 1968.

95. Norman Borlaug Interview Archived May 13, 2013, at the Wayback Machine., Academy of Achievement, May 12, 2008
96. Easterbrook, G. 1997. Forgotten Benefactor of Humanity. The Atlantic Monthly.
97. Borlaug, N. E. 1972. Nobel Lecture, December 11, 1970. From Nobel Lectures, Peace 1951–1970, Frederick W. Haberman Ed., Elsevier Publishing Company, Amsterdam
98. "Nobel Lecture — The Nobel Peace Prize 1970 Norman Borlaug". nobelprize.org.
99. Angelsen, A., and D. Kaimowitz. 2001. "The Role of Agricultural Technologies in Tropical Deforestation"."Agricultural Technologies and Tropical Deforestation" (PDF). Archived from the original (PDF) on 2005-09-29. Retrieved 2005-07-17. . CABI Publishing, New York
100. Borlaug, Norman; Garrett, Peter (December 18, 1999). "Between the Tynes / Chronicles of the Future — Program 6 Earth, wind & fire". The Weekend Australian.
101. Leonard, Andrew (July 16, 2007). "Show organic farmers the money". Salon.com.
102. Cockburn, Alexander (June 29, 2003). "Corporate Interests Keep World's Poor Hungry". Sunday Business Post. Archived from the original on January 12, 2010. Retrieved 27 May 2015.
103. Billions served. Interview with Reason Magazine. April 2000
104. Herbert Hoover Presidential Library and Museum. 2002.Four Iowans Who Fed The World, Norman Borlaug: Geneticist at the Wayback Machine (archive index)
105. Tierney, John (May 19, 2008). "Greens and Hunger". TierneyLab — Putting Ideas in Science to the Test. New York Times. Retrieved 2009-02-13.
106. Press, Robert. "Borlaug: sowing 'Green Revolution' among African leaders". Christian Science Monitor. Retrieved 2009-09-06.
107. The Carter Center. "Norman Borlaug, Senior Consultant in Agriculture". Retrieved 2008-07-17.
108. The Carter Center. "The Carter Center Agriculture Program". Retrieved 2008-07-17.
109. The Carter Center (2007-05-02). "Exhibit to Highlight Progress For Peace, Health, Human Rights". Retrieved 2008-07-17.
110. The Murugappa Group. 2005. Food for Thought
111. Norman Borlaug: genetic modification can feed the world, Chron.com, July 13, 2008
112. Borlaug, N.E. (2000), "Ending world hunger: the promise of biotechnology and the threat of antiscience zealotry", Plant Physiology, 124 (2): 487–490, doi:10.1104/pp.124.2.487, PMC 1539278 , PMID 11027697
113. Rozwadowski, Kevin; Kagale, Sateesh (nd), Global Food Security: The Role of Agricultural Biotechnology Commentary (PDF), Saskatoon, Saskatchewan:

Saskatoon Research Centre, Agriculture and Agri-Food Canada, retrieved 12 January 2014

114. Conko, Greg. The Man Who Fed the World. Openmarket.org. September 13, 2009.
115. "Population Media Center 2008 Annual Report" (PDF). Populationmedia.org. 2008.
116. "Nobel Prize-winning scientist Norman Borlaug, father of the 'green revolution,' dies at age 95".
117. Associated Press in Dallas (2009-09-13). "Norman Borlaug, the Nobel winner who fed the world, dies aged 95". London: The Guardian. Retrieved 2009-09-15.
118. "PM pays tribute to Father of Green Revolution Borlaug". Rediff. September 14, 2009. Retrieved 2009-09-15.
119. "UN food agency pays tribute to 'father' of Green Revolution". United Nations. 14 September 2009. Retrieved 2009-09-15.
120. "Tributes to Dr. Norman E. Borlaug from around the world". World Food Prize. 2009-09-14. Retrieved 2009-09-15.
121. "Nobel Peace Prize 1970 — Presentation Speech".
122. "List of Distinguished Fellows of the Iowa Academy of Science".
123. http://www.jeffersonawards.org/pastwinners/national
124. "SAHF Inductees". hostfest.com. Norsk Høstfest.
125. "Fellowship of the Royal Society 1660-2015". London: Royal Society. Archived from the original on 2015-07-15.
126. iCar.org
127. "Public Welfare Award". National Academy of Sciences. Archived from the original on 4 June 2011. Retrieved 18 February 2011.
128. "Biographical background on 2005 Dartmouth honorary degree recipients NORMAN E. BORLAUG (Doctor of Science)". www.dartmouth.edu. Retrieved 2009-08-24.
129. Dr. Norman E. Borlaug's Curriculum Vitae
130. "Father of India's Green Revolution" given Padma VibhushanRediff.com, August 24, 2006.
131. Bjordal, J. Cathedral Peace Window honors Dr Norman Borlaug and Jimmy Carter Journal of the American Chestnut Foundation, vol. 18 no.2 Fall 2004, p. 9. Retrieved 2009-09-06.
132. "Eat This!". Penn & Teller: Bullshit!. Season 1. Episode 11. April 4, 2003.
133. An Act To award a congressional gold medal to Dr. Norman E. Borlaug
134. "The Dr. Norman Borlaug Bronze Medal". The United States Mint. Retrieved 2011-02-16.

135. Alumnus Norman Borlaug receives National Medal of Science
136. List of Fellows of Bangladesh Academy of Sciences Archived April 15, 2010, at the Wayback Machine.

CHAPTER 3

The Art of Arboriculture

Chapter Contents

- **Introduction**
- **Fruit Tree Pruning**
- **Tree Shaping**
- **Bonsai**
- **Tree Fork**

Arboriculture is the science art of cultivating trees. This includes selecting tress for specific purposes, planting them, and caring for them. In many cases, arboriculture is applied in garden settings, in which a person selects trees that are pest resistant and most likely to adapt to the location. The art of arboriculture is used by the gardener to create a tree display that is aesthetically pleasing, ornamental, and long lasting.

INTRODUCTION

Arboriculture is the cultivation, management, and study of individual trees, shrubs, vines, and other perennial woody plants. The science of arboriculture studies how these plants grow and respond to cultural practices and to their environment. The practice of arboriculture includes cultural techniques such as tree selection, planting, training, fertilization, pest and pathogen control, pruning, shaping, and removal of unnecessary tree parts.

A person who practices or studies arboriculture can be termed an 'arborist' or an 'arboriculturist'. Risk management, legal issues, and aesthetic considerations have come to play prominent roles in the practice of arboriculture. Businesses often need to hire arboriculturists to complete "tree hazard surveys" and generally manage the trees on-site to fulfill occupational safety and health obligations.

Arboriculture is primarily focused on individual woody plants and trees maintained for permanent landscape and amenity purposes, usually in gardens, parks or other populated settings, for the enjoyment, protection, and benefit of human beings.

An arborist practicing arboriculture: using a chainsaw to cut a eucalyptus tree in a park at Kallista, Victoria.

FRUIT TREE PRUNING

Fruit tree pruning is the cutting and removing of selected parts of a fruit tree. It spans a number of horticultural techniques. Pruning often means cutting branches back, sometimes removing smaller limbs entirely. It may also mean removal of young shoots, buds, and leaves.

Established orchard practice of both organic and non-organic types typically includes pruning. Pruning can control growth, remove dead or diseased wood, and stimulate the formation of flowers and fruit buds. It is widely stated that careful attention to pruning and training young trees improve their later productivity and longevity, and that good pruning and training can also prevent injury from weak crotches or forks (where a tree trunk splits into two or more branches) that break from the weight of fruit, snow, or ice on the branches.

Some sustainable agriculture or permaculture personalities, such as Sepp Holzer and Masanobu Fukuoka, advocate and practice no-pruning methods, which runs counter to the widespread confidence in the idea that pruning produces superior results compared with not pruning. Many books about fruit-growing assert advantages and disadvantages of pruning or not pruning, although without randomized controlled trials, it is hard to separate theorizing and traditional knowledge from evidence-based recommendations.

A community apple orchard originally planted for productive use during the 1920s, in Westcliff on Sea (Essex, England), illustrating long neglected trees that have recently been pruned to renew their health and cropping potential

Overview

An apple tree sprout is being converted to a branched, fruit-bearing spur by an arborist. Numbers show the sequence of cuts, which occurred during two years.

Plants form new tissue in an area called the meristem, located near the tips of roots and shoots, where active cell division takes place. Meristem growth is aimed at ensuring that leaves are quickly elevated into sunlight, and that roots are able to penetrate deeply into the soil. Once adequate height and length is achieved by the stems and roots, they begin to thicken to support the plant. On the shoots, these growing tips of the plant are called *apical* buds. The *apical meristem* (or tip) produces the growth hormone auxin, which not only promotes cell division, but also diffuses downwards and inhibits the development of lateral bud growth that otherwise competes with the apical tip for light and nutrients. Removing the apical tip and its suppressive hormone lets lower, dormant lateral buds develop, and the buds between the leaf stalk and stem produce new shoots that compete to become lead growth. Manipulating this natural response to damage (known as the principle of apical dominance) by processes such as pruning (as well as coppicing and pollarding) allows the arborist to determine the shape, size, and productivity of many fruiting trees and bushes. The main aim when pruning fruit trees is usually to maximize fruit yield. Unpruned trees tend to produce large numbers of small fruits that may be difficult to reach when harvesting by hand. Branches can become broken by the weight of the crop, and the cropping may become biennial (that is, bearing fruit only every other year). Overpruned trees on the other hand tend to produce light crops of large, flavourless fruit that does not store well. Careful pruning balances shoot growth and fruit production.

One of the simplest instructions given in nearly every article or book chapter on the subject is that no branches should cross each other, that is, rub against each other, and that one of them should be selected and removed.

Formative Pruning of Bush Trees

In the early years of the tree's life, it is important to develop a framework sufficiently strong to bear the weight of crops. This requires formative pruning to reinforce the tree. Formative pruning of apple (*Malus pumila*) and pear (*Pyrus communis*) trees should be carried out in the dormant winter months. For the Northern hemisphere, this should occur between November and March; For the Southern hemisphere, June and September. Stone fruits—such as cherries, plums, and gages—have different requirements, and should not be pruned in dormant months.

One-year-old tree

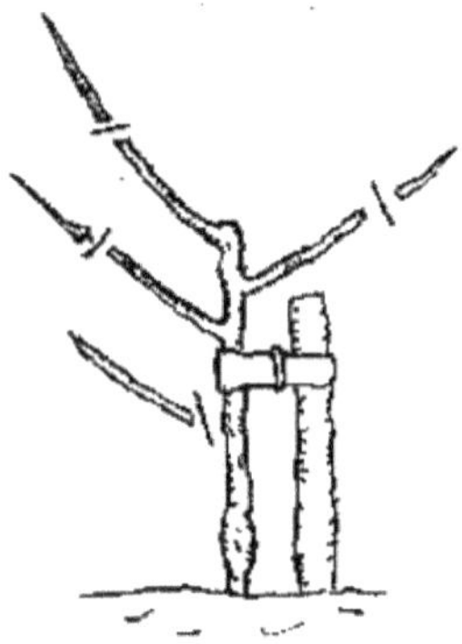

Two-year-old tree

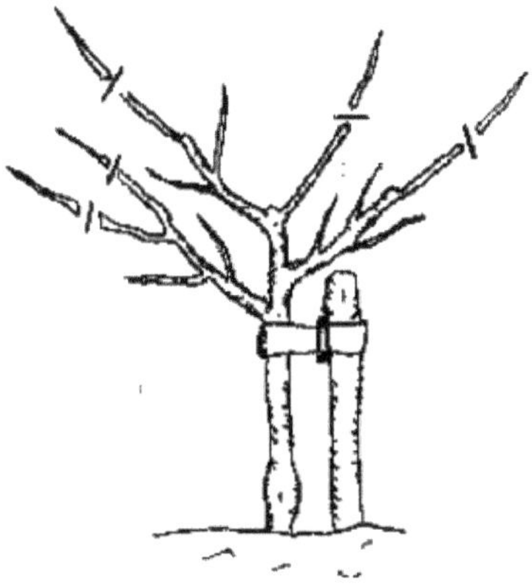

Three-year-old tree

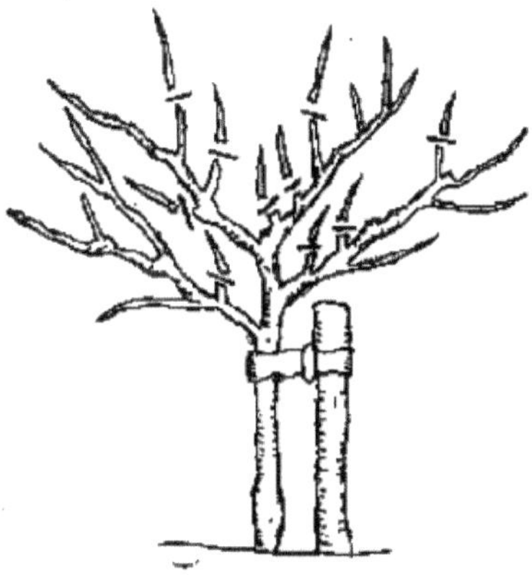

Four-year-old tree

Maiden Tree

A maiden whip (a one-year-old tree with no side shoots) should be pruned to a bud with two buds below it at about 80 cm from the ground immediately after planting to produce primary branches during the first growing season. A feathered maiden (that is, a one-year-old tree with several side branches) should have its main stem pruned back to three or four strong shoots at 80 cm from the ground. Side shoots should be shortened by two thirds of their length to an upward or outward facing bud. Lower shoots should be removed flush with the stem.

Two Year

Remove lower shoots and prune between three and five of the best-placed shoots by half to an upwards or outwards facing bud to form what becomes the tree's main structural branches. Remove any inward-facing shoots.

Three Year

Prune leading shoots of branches selected to extend the framework by half, to a bud facing in the desired direction. Select four good laterals to fill the framework and shorten these by a half. Prune any remaining laterals to four buds to form fruiting spurs.

Four Year

The tree has begun to fruit and requires only limited formative pruning. Shorten leaders by one third and prune laterals not required to extend the framework to four buds.

Five Year and Onwards

The tree is established, and should be pruned annually as described in the following section.

Pruning the Cropping Tree

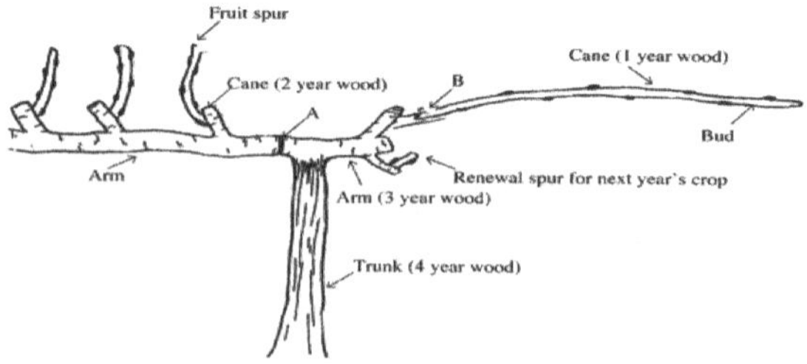

Spur pruning

Before pruning, distinguish between spur-bearing varieties, tip-bearing varieties, and an intermediate between the two that bears both on spurs and at the tips. Spur-bearing trees occur more frequently than tip-bearing trees, and they bear most of their fruit yearly at the end of short lateral pieces of wood (spurs) up to about 4 inches long.

Spur-bearing types include apples of the varieties 'Cox's Orange Pippin', 'James Grieve' and 'Sunset', and pears such as 'Conference', 'Doyenne du Commice', and 'Williams Bon Chretien'. Tip-bearers on the other hand produce most of their fruit buds at the tips of slender shoots grown the previous summer, and include the apples 'Worcester Pearmain' and 'Irish Peach', and the pears such as 'Jargonelle' and 'Josephine de Malines'. There are basically three types of pruning that are applied once the main shape of the tree has been established. These are:

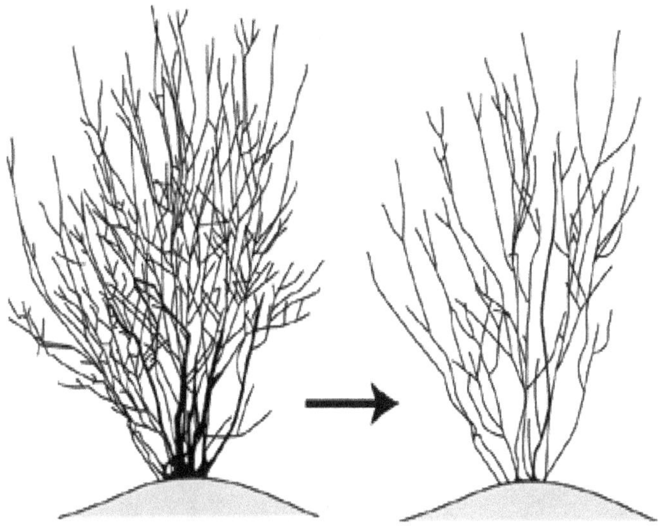

Renewal pruning

- **Spur pruning**: Spur bearing varieties form spurs naturally, but spur growth can also be induced.
- **Renewal pruning**: This also depends on the tendency of many apple and pear trees to form flower buds on unpruned two-year-old laterals. It is a technique best used for the strong laterals on the outer part of the tree where there is room for such growth. Pruning long-neglected fruit trees is a task that should be undertaken over a lengthy period, with not more than one third of the branches that require removal being taken each year.
- **Regulatory pruning**: This is carried out on the tree as a whole, and is aimed at keeping the tree and its environment healthy, e.g., by keeping the centre open so that air can circulate; removing dead or diseased wood; preventing branches from becoming overcrowded (branches should be roughly 50 cm apart and spurs not less than 25 cm apart along the branch framework); and preventing any branches from crossing.

Pruning of Tip Bearers

Tip-bearers should be pruned lightly in winter using the regulatory system (see above). Any maiden shoots less than 25 cm in length should be left untouched as they have fruit buds at their tips. Longer shoots are spur pruned to prevent overcrowding and to stimulate the production of more short-tip-bearing shoots the following year. Branch leaders are 'tipped', removing the top three or four buds to a bud facing in the desired direction to make them branch out and so produce more tip-bearing shoots.

TREE SHAPING

Tree shaping (also known by several other alternative names) uses living trees and other woody plants as the medium to create structures and art. There are a few different methods used by the various artists to shape their trees, which share a common heritage with other artistic horticultural and agricultural practices, such as pleaching, bonsai, espalier, and topiary, and employing some similar techniques. Most artists use grafting to deliberately induce the inosculation of living trunks, branches, and roots, into artistic designs or functional structures.

Tree shaping has been practiced for at least several hundred years, as demonstrated by the living root bridges built and maintained by the Khasi people of India. Early 20th century practitioners and artisans included banker John Krubsack, Axel Erlandson with his famous circus trees, and landscape engineer Arthur Wiechula. Contemporary designers include "Pooktre" artists Peter Cook and Becky Northey, "arborsculpture" artist Richard Reames, and furniture designer Dr Chris Cattle, who grows "grownup furniture".

Needle & Thread Tree by Axel Erlandson

Artist Peter Cook seated in a living garden chair grown via the Pooktre method

Methods

There are various methods of shaping a tree. Some of these processes are still experimental, whereas others are still in the research stage. These methods use a variety of horticultural, arboricultural, techniques to achieve an intended design. Chairs, tables, living spaces and art may be shaped from growing trees. Some techniques used are unique to a particular practice, whereas other techniques are common to all, though the implementation may be for different reasons. These methods usually start with an idea of the intended outcome. Some practitioners start with detailed drawings, or designs, other artists start with what the tree already has. Each process has its own time frame and a different level of involvement from the tree shaper. The trees might then either remain growing, as with the living Pooktre garden chair, or perhaps be harvested as a finished work like John Krubsack`s chair.

Aeroponic Culture

Chair created using aeroponic root shaping

The oldest known living examples of woody plant shaping are the aeroponically cultured living root bridges built by the ancient War-Khasi people of the Cherrapunjee region in India. These are being maintained and further developed today by the people of that region. Aeroponic growing was first formally studied by W. Carter in 1942. Carter researched air culture growing and described "a method of growing plants in water vapor to facilitate examination of roots". Later researchers, including L. J Klotz and G. G. Trowel, expanded on his work. In 1957, F. W. Went described "the process of growing plants with air-suspended roots and applying a nutrient mist to the root section," and in it he coined the word 'aeroponics' to describe that process. In 2008, root researcher and craftsman Ezekiel Golan described and secured a patent for a process which allows the roots of some aeroponically grown woody plants to lengthen and thicken while still remaining flexible. At lengths of perhaps 6 metres (20 ft) or more, the soft roots can be formed into pre-determined shapes which will continue thickening after the shapes are formed and as they continue to grow. Newer techniques and applications, such as eco-architecture, may allow architects to design, grow, and form large permanent structures, such as homes, by shaping aeroponically grown plants and their roots.

Instant Tree Shaping

Arborsculpture bench by Richard Reames created using the techniques as described in his books *How to grow a chair* and *Arborsculpture*

Instant tree shaping starts with mature trees, perhaps 6–12 ft. (2–3.6 m) long and 3-4in (7.6–10 cm) in trunk diameter, which are bent and woven into the desired design and held until cast. Understanding a tree's fluid dynamics is important to achieving the desired result.

Bending is sometimes used to achieve a design. If a plant's tissue is bent at too sharp an angle it may break, which can be mostly avoided by un-localizing the bend. This is achieved by making small bends along the curve of the tree. Bends are then held in place for several years until their form is permanently cast. The tree's rate of growth determines the time necessary to overcome its resistance to the initial bending. The work of bending and securing in this way might be accomplished in an hour or perhaps in an afternoon depending on the design.

Ring barking is sometimes employed to help balance a design by slowing the growth of too-vigorous branches or stopping the growth of inopportunely placed branches, using different degrees of ring barking, from simple scoring to complete removal of a 3/8"-wide (1 cm) band of bark.

Creasing is folding trees such as willow and poplar over upon themselves, creating a right angle. This method is more radical than bending.

With this method it is possible to perform initial bending and grafting on a project in an hour, as with *Peace in Cherry* by Richard Reames, removing supports in as little as a year and following up with minimal pruning thereafter.

Gradual Tree Shaping

"Grownup furniture" by Chris Cattle created using a gradual tree shaping method

Gradual tree shaping starts with designing and framing. These are fundamental to the success of the piece. Once these are set up, young seedlings or saplings 3–12 in. (7.6–30.5 cm) long are planted.

The training starts with young seedlings, saplings or the stems of trees when they are very young, which are gradually shaped while the tree is growing to form the desired shape. There is a small area that just behind the growing tip that forms the final shape. The shaping zone It is the shaping of this area requires day to day or weekly guiding of the new growth. The growth is guided along predetermined design pathways; this may be a wooden jig or complex wire design.

With this method the time frame is longer than the other methods. A chair design might take 8 to 10 years to reach maturity Some of Axel Erlandson trees`s took as long as 40 years to assume their finished shapes.

Common Techniques

Framing

Framing may be used for various purposes and might consist of any one or a combination of several materials, such as timber, steel, tubes made of hollow out trees, complex wire designs, wooden jigs, or the tree itself living :178 or dead.:58It can be used in many project designs to support grafted joints until the grafts are well-established. Some process might employ framing to hold a shape created by bending or fletching mature trees until the tissues have overcome their resistance to the initial bending and grown enough annual rings to cast the design permanently. Others might use framing to support and shape the growth of young saplings until they are strong enough to maintain an intended shape without support. Still other approaches might employ frames to guide the roots of aeroponically grown trees into desired shapes.

Grafting

Grafting is a common technique used by all the different methods. Grafting exploits the natural biological process of inosculation. Grafting is where a branch or plant is cut and a piece of another plant is added and held in place. There are various types of grafting, in all types the idea is to encourage the tissues of one plant to fuse with those of another.

Grafting is applied to create permanent connections and joins. In some cases the trees are grafted while they are growing in others the mature trees may be intertwining and then grafting together the stems of two or more trees in order to create chairs, ladders, and other fanciful sculptures.

Pruning

Pruning can be used to balance a design by controlling and directing growth into a desired shape. Pruning above a leaf node can steer plant growth in the direction of the natural placement of that leaf bud. Pruning may also be used to keep a design free of unwanted branches and to reduce canopy size. Pruning is sometimes the only technique used to craft a project. Deciduous trees are mainly pruned in winter, while they are dormant above-ground, although sometimes it is necessary to prune them during the growing season. Trees repeatedly subjected to hard pruning may experience stunted growth, and some trees may not survive this treatment.

Structure

Living grown structures have a number of structural mechanical advantages over those constructed of lumber and are more resistant to decay. While there are some decay organisms that can rot live wood from the outside, and though living trees can carry decayed and decaying heartwood inside them; in general, living trees decay from the inside out and dead wood decays from the outside in. Living wood tissue, particularly sapwood, wields a very potent defense against decay from either direction, known as compartmentalization. This protection applies to living trees only and varies among species.

Growing structures is not as easy as it would seem. Quick growing willows have been used to grow building structures, they provide support or protection. A young group of German architects are in the process of growing such structure and they are continually monitored and checked. Once the trees are of age to be able to take on load-bearing weight they are tested for stability and strength by a structural engineer. Once this is approved the supporting framework is removed. Projects are limited to the trees` weight loading ability and growth. This is being studied and the load capacity will be proved by testing on prototypes.

Design Options

Becky's Mirror **by Pooktre**

Designs may include abstract, symbolic, or functional elements. Some shapes crafted and grown are purely artistic; perhaps cubes, circles, or letters of an alphabet, while other designs might yield any of a wide variety of useful shapes, such as clothes hangers, laundry and wastepaper bins, ladders,furniture, tools, and tool handles. Eye-catching structures such as living fences and jungle gyms can also be grown, and even large architectural designs such as live archways, domes, gazebos,tunnels, and theoretically entire homes are possible with careful planning, planting, and culturing over time. The Human Ecology Design team (H.E.D.) at the Massachusetts Institute of Technology is designing homes that can be grown from native trees in a variety of climates.

Suitable trees are installed according to design specifications and then cultured over time into intended structures. Some designs may use only living, growing wood to form the structures, while others might also incorporate inclusions such as glass, mirror, steel and stone, any of which might be used either as either structural or aesthetic elements. Inclusions can be positioned in a project as it is grown and, depending on the design, may either be removed when no longer needed for support or left in place to become fixed inclusions in the growing tissue.

Chronology of Notable Practitioners

War-Khasi people

The ancient War-Khasi people of India worked with the aerial roots of native banyan fig trees, adapting them to create footbridges over watercourses. Modern people of the Cherrapunjee region carry on this traditional building craft. Roots selected for bridge spans are supported and guided in darkness as they are being formed, by threading long, thin, supple banyan roots through tubes made from hollowed-out trunks of woody grasses. Preferred species for the tubes are either bamboo or areca palm, or ‹kwai` in Khasi, which they cultivate for areca nuts. The Khasi incorporate aerial roots from overhanging trees to form support spans and safety handrails. Some bridges can carry fifty or more people at once. At least one example, over the Umshiang stream, is a double-decker bridge. They can take ten to fifteen years to become fully functional and are expected to last up to 600 years.

John Krubsack

John Krubsack was an American banker and farmer from Embarrass, Wisconsin. He shaped and grafted the first known grown chair, harvesting it in 1914. He lived from 1858 to 1941. He had studied tree grafting and become a skilled found-wood furniture crafter. The idea first came to him to grow his own chair during a weekend wood-hunting excursion with his son. He started box elder seeds in 1903, selecting and planting either 28 or 32 of the saplings in a carefully designed pattern in the spring of 1907. In the spring of 1908, the trees had grown to six feet tall and he began training them along a trellis, grafting the branches at critical points to form the parts of his chair. In 1913, he cut all the trees except those forming the legs, which he left to grow and increase in diameter for another year, before harvesting and drying the chair in 1914; eleven years after he started the box elder seeds. Dubbed *The Chair that Lived*; it is the only known tree shaping that John Krubsack did. The chair is on permanent display in a Plexiglas case at the entrance of Noritage Furniture; the furniture manufacturing business now owned by Krubsack`s descendants, Steve and Dennis Krubsack.

Axel Erlandson

Basket Tree by Axel Erlandson

Needle & Thread Tree by Axel Erlandson

Axel Erlandson was a Swedish American farmer who started training trees as a hobby on his farm in Hilmar, California, in 1925. He was inspired by observing a natural sycamore inosculation in his hedgerow. In 1945, he moved his family and the best of his trees from Hilmar to Scotts Valley, California and in 1947, opened an horticultural attraction called the Tree Circus.

Erlandson lived from 1884 to 1964; training more than 70 trees during his lifetime. He considered his methods trade secrets and when asked how he made his trees do this, he would only reply, "I talk to them." His work appeared in the column of *Ripley's Believe It or Not!* twelve times. 24 trees from his original garden have survived transplanting to their permanent home at Gilroy Gardens in Gilroy, California. His *Telephone Booth Tree* is on permanent display at the American Visionary Art Museum in Baltimore, Maryland and his *Birch Loop* tree is on permanent display at the Museum of Art and History in Santa Cruz, California. Both of these are preserved dead specimens.

Arthur Wiechula

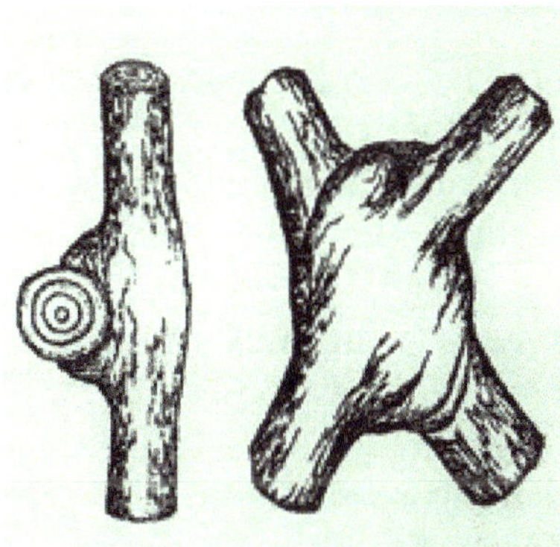

19th-century sketch by Arthur Wiechula of inosculated branches

Arthur Wiechula was a German landscape engineer who lived from 1868 to 1941. In 1926, he published *Wachsende Häuser aus lebenden Bäumen entstehend* (Developing Houses from Living Trees) in German. In it, he gave detailed illustrated descriptions of houses grown from trees and described simple building techniques involving guided grafting together of live branches; including a system of v-shaped lateral cuts used to bend and curve individual trunks and branches in the direction of a design, with reaction wood soon closing the wounds to hold the curves. He proposed growing wood so that it constituted walls during growth, thereby enabling the use of young wood for building. Weichula never built a living home, but he grew a 394` wall of Canadian poplars to help keep the snow off of a section of train tracks.

Dan Ladd

Dan Ladd is a Northampton, Massachusetts based American artist who works with trees and gourds. He began experimenting with glass, china, and metal inclusions in trees in 1977 in Vermont and started planting trees for Extreme Nature in 1978. He became inspired by inosculation he noticed in nature and by the growth of tree trunks around man-made objects such as fences and idle farm equipment. He shapes and grafts trees, including their fruits and their roots, into architectural and geometric forms. Ladd calls human-initiated inosculation 'pleaching` and calls his own work ‹tree sculpture`. Ladd binds a variety of objects to trees, for live wood to grow around and be incorporated, including teacups, bicycle wheels, headstones, steel spheres, water piping, and electrical conduit. He guides roots into shapes, such as stairs, using above-ground wooden and concrete forms and even shapes woody, hard-shelled Lagenaria gourds by allowing them to grow into detailed molds. A current project at the DeCordova and Dana Museum and Sculpture Park in Lincoln, Massachusetts incorporates eleven American Liberty Elm trees grafted next to each other to form a long hillside stair banister. Another of his installations, *Three Arches*, consists of three pairs of 14-foot sycamore trees, which he grafted into arches to frame different city views, at Frank Curto Park in Pittsburgh.

Nirandr Boonnetr

Nirandr Boonnetr is a Thai furniture designer and crafter. He became inspired as a child, both by a photograph of some unusually twisted coconut palms in southern Thailand and by a living fallen tree he noticed, which had grown new branches along its trunk, forming a kind of canopied bridge. His hobby began in 1980 because of his concern the Thailand forests are being ravaged by woodcarvers to the point that one day the industry would eventually carve itself out of existence. He began his first piece, a guava chair, around 1983. Originally intended as something for his children to climb and play on, the piece evolved into a living tree chair.[91] In fifteen years he created six pieces of "living furniture,"including five chairs and a table. The Bangkok Post dubbed him the father of Living Furniture. Shortly thereafter, he presented a chair as a gift to her Royal Highness, Princess Sirindhorn. Nirandr Boonnetr has written a detailed, step-by-step booklet of instructions hoping his hobby of living furniture will spread to other countries. One of his chairs was exhibited in the Growing Village pavilion at the World`s Fair Expo 2005 in Nagakute, Aichi, Japan.

Peter Cook and Becky Northey

Peter Cook and Becky Northey are Australian artists who live in South East Queensland. Peter Cook became inspired to grow a chair in 1987, after visiting three figs trees in a remote corner of his property. He started the next day, with 7 willow cuttings. In 1988, he planted a wattle intended for harvest as a potted plant stand. Becky Northey moved to Peter`s property in 1995 and the two formed Pooktre.

Their methods involve guiding a tree`s growth along predetermined wired design pathways over a period of time. They shape growing trees both for living outdoor art and for intentional harvest. They most often use Myrobalan Plum for shaping. Examples of their functional artwork include a growing garden table, a harvested coffee table, hat stands, mirrors, and a gemstone neck piece.

People trees, by Pooktre

Peter and Becky exhibited eight of their creations, including two people trees. in the Growing Village pavilion at the World`s Fair Expo 2005 in Nagakute, Aichi Prefecture, Japan. Their work was published in the annual book series, *Ripley's Believe It or Not*.

Richard Reames

Richard Reames's *Peace in Cherry*

Richard Reames is an American nurseryman and author based in Williams, Oregon, where he owns and manages a nursery, and design studio collectively named Arborsmith Studios. He was inspired by the works of Axel Erlandson,[:150:16] and began sculpting trees in 1991 or 1992. He began his first experimental grown chairs [:57] in the spring of 1993.[:85]

In 1995, Reames wrote and published his first book, *How to Grow a Chair: The Art of Tree Trunk Topiary.* In it, he coined the word arborsculpture. In 2005, he published his second book, *Arborsculpture: Solutions for a Small Planet.* He has lectured in Australia and gives live demonstrations of bending and weaving a chair at garden shows, fairs and folk art festivals around America.

Christopher Cattle

Christopher Cattle's grown stool in sycamore

Christopher Cattle is a retired furniture design professor from England. He started his first planting of furniture in 1996. According to Cattle, he developed an idea to train and graft trees to grow into shapes, which came to him in the late 1970s, in response to questions from students asking how to build furniture using less energy. Using various species of trees and wooden jigs to shape them, he has grown 15 three-legged stools to completion.

Cattle has multiple plantings in at least four different locations in England. He participates in woodland and craft shows in England and at the Big Tent at Falkland Palace in Scotland. He exhibited his grown stools at the World's Fair Expo 2005 in the Growing Village pavilion at Nagakute, Japan. He aims to encourage as many people as possible to grow their own furniture, and envisions that, "One day, furniture factories could be replaced by furniture orchards." Cattle calls his works grown up furniture and grown stools, but also refers to them as grown furniture, calling them "the result of mature thinking."

Mr. Wu

Mr. Wu is a Chinese pensioner who designs and crafts furniture in Shenyang, Liaoning, China. He has patented his technique of growing wooden chairs , had designed, grown, and harvested one chair, and had six more growing in his garden. Wu uses young elm trees, which he says are pliant and do not break easily. He also says that it takes him about five years to grow a tree chair.

Related Practices

Other artistic horticultural practices such as bonsai, espalier, and topiary share some elements and a common heritage, though a number of distinctions may be identified.

Bonsai

Bonsai is the art of growing trees in small containers. Bonsai uses techniques such as pruning, root reduction, and shaping branches and roots to produce small trees that mimic, full-sized mature trees. Bonsai is not intended for production of food, but instead mainly for contemplation by viewers, like most fine art.

Espalier

Espalier is the art and horticultural practice of training tree branches onto ornamental shapes along a frame for ascetic and fruit production by grafting, shaping and pruning the branches so that they grow flat, frequently in formal patterns, against a structure such as a wall, fence, or trellis. The practice is commonly used to accelerate and increase production in fruit-bearing trees and also to decorate flat exterior walls while conserving space.

Pleaching

Pleaching is a technique of weaving the branches of trees into a hedge commonly, deciduous trees are planted in lines, then pleached to form a flat plane on clear stems above the ground level. Branches are woven together and lightly tied. Branches in close contact may grow together, due to a natural phenomenon called inosculation, a natural graft. Pleach also means weaving of thin, whippy stems of trees to form a basketry affect.

Topiary

Topiary is the horticultural practice of shaping live trees, by clipping the foliage and twigs of trees and shrubs to develop and maintain clearly defined shapes, often geometric or fanciful. The hedge is a simple form of topiary used to create boundaries, walls or screens. Topiary always involves regular shearing and shaping of foliage to maintain the shape.

Plantings for the Future

The Fab Tree Hab

Fab Tree Hab 3D render

Three MIT designers – Mitchell Joachim, Lara Greden and Javier Arbona – created a concept of a living tree house which nourishes its inhabitants and merges with its environment. The project of Fab Tree Hab is expected to take a minimum of five years to grow the home. The plans are for the interior to be lined with clay and plastered to keep the weather outside and to look normal. The exterior is to be all natural.

The Patient Gardener

A Swedish architectural firm VisionDivision took part in a week-long workshop at the Italian university Politecnico di Milano with the students. The result was an 80-year plan of a living cheery tree dome in an hourglass shape and grown furniture. Framing for the dome, table and a lawn chair were built. Ten Japanese cherry trees were planted in a diameter of eight-meter circle. Four of these trees are to be living staircases to a future top level. The stair trees will have their branches grafted into each other to form the rungs. VisionDivision`s architects helped the students and instructors to create an easy maintenance plan for future gardeners of the university.

Baubotanik Tower

Ferdinand Ludwig designed this tower as part of his doctoral thesis with the help of Prof. Dr. Speck. "Speck become the botanical co-supervisor" said Ferdinand. Growing at the University of Stuttgart is a three-storey tower of living white willows (*Salix alba*).

This nine-meter-tall construction is almost fully grown, with a base area of around eight square meters. [:86]

The framing is made up of mainly steel scaffolding which is supporting the growing trees, while keeping them to the correct form. They started with 400 white willow (Salix alba) grown in baskets on multiple levels with one row of willows planted into the ground. Once the trees were two meters tall, they were planted at the different levels of the tower. These plants are then trained to the design.

The root system of the bottom level of willows needs to develop large enough to support the willows on the above levels, so that the scaffold becomes obsolete and then it and the watering and fertilising baskets can be removed altogether. [:86]

The trees are grafted together with the objective of all the different plants eventually becoming a single organism. The overall aim is to have a living structure with the strength to support itself and to carry a working load. Ferdinand predicts the tower will be stable enough to support itself in five to ten years. Ferdinand does state "However, these are only estimates."

Alternative Names

The practice of shaping living trees has several names. Practitioners may have their own name for their techniques, so a standard name for the various practices has not emerged. Richard Reames calls the practice "arborsculpture"; Dan Ladd calls his work "tree sculpture";Nirandr Boonnetr's work is called "living furniture"; Christopher Cattle calls his works "grown up furniture" and "grown stools"; while Peter Cook and Becky Northey call their work "Pooktre".

The following names are also encountered:

- Arbortecture
- Biotecture/Biotechture
- Grown furniture
- Living Art
- Pleaching
- Tree training

In Fiction and Art

In 1516, Jean Perréal painted an allegorical image, *La complainte de nature à l'alchimiste errant*, (*The Lament of Nature to the Wandering Alchemist*), in which a winged figure with arms crossed, representing nature, sits on a tree stump with a fire burning in its base, conversing with an alchemist in an ankle length coat, standing outside of his stone-laid shoreline laboratory. Live resprouting shoots emerge from either side of the tree stump seat to form a fancifully twined and inosculated two-story-tall chair back.

In 1758, Swedish scientist, philosopher, Christian mystic, and theologian Emanuel Swedenborg published *Earths in the Universe*, in which he wrote of visiting another planet where the residents dwelled in living groves of trees, whose growth they had planned and directed from a very young stage into living quarters and sanctuaries.

In the late 19th century, Styrian Christian mystic and visionary Jakob Lorber published *The Household of God*. In it, he wrote about the wisdom of planting trees in a circle, because once grown together, the ring of trees would be a much better house than could be built.

BONSAI

Bonsai is a Japanese art form using trees grown in containers. Similar practices exist in other cultures, including the Chinese tradition of *penjing* from which the art originated, and the miniature living landscapes of Vietnamese *hòn non b* . The Japanese tradition dates back over a thousand years.

"Bonsai" is a Japanese pronunciation of the earlier Chinese term *penzai*. The word *bonsai* is often used in English as an umbrella term for all miniature trees in containers or pots. This article focuses on bonsai as defined in the Japanese tradition.

The purposes of bonsai are primarily contemplation (for the viewer) and the pleasant exercise of effort and ingenuity (for the grower). By contrast with other plant cultivation practices, bonsai is not intended for production of food or for medicine. Instead, bonsai practice focuses on long-term cultivation and shaping of one or more small trees growing in a container.

A bonsai is created beginning with a specimen of source material. This may be a cutting, seedling, or small tree of a species suitable for bonsai development. Bonsai can be created from nearly any perennial woody-stemmed tree or shrub species that produces true branches and can be cultivated to remain small through pot confinement with crown and root pruning. Some species are popular as bonsai material because they have characteristics, such as small leaves or needles, that make them appropriate for the compact visual scope of bonsai.

The source specimen is shaped to be relatively small and to meet the aesthetic standards of bonsai. When the candidate bonsai nears its planned final size it is planted in a display pot, usually one designed for bonsai display in one of a few accepted shapes and proportions. From that point forward, its growth is restricted by the pot environment. Throughout the year, the bonsai is shaped to limit growth, redistribute foliar vigor to areas requiring further development, and meet the artist's detailed design.

The practice of bonsai is sometimes confused with dwarfing, but dwarfing generally refers to research, discovery, or creation of plant cultivars that are permanent, genetic miniatures of existing species. Bonsai does not require genetically dwarfed trees, but rather depends on growing small trees from regular stock and seeds. Bonsai uses cultivation techniques like pruning, root reduction, potting, defoliation, and grafting to produce small trees that mimic the shape and style of mature, full-size trees.

Japanese white pine from the National Bonsai & Penjing Museum at the United States National Arboretum.

Bonsai at the Omiya Bonsai Art Museum

Bonsai at the National Bonsai & Penjing Museum at the United States National Arboretum

Early Versions

The earliest illustration of a penjing is found in the Qianling Mausoleum murals at the Tang-dynasty tomb of Crown Prince Zhanghuai, dating to 706.

The Japanese art of bonsai originated from the Chinese practice of penjing. From the 6th century onward, Imperial embassy personnel and Buddhist students from Japan visited and returned from mainland China. They brought back many Chinese ideas and goods, including container plantings. Over time, these container plantings began to appear in Japanese writings and representative art.

In the medieval period, recognizable bonsai were portrayed in handscroll paintings like the *Ippen shonin eden* (1299). The 1195 scroll *Saigyo Monogatari Emaki* was the earliest known to depict dwarfed potted trees in Japan. Wooden tray and dish-like pots with dwarf landscapes on modern-looking wooden shelves also appear in the 1309 *Kasuga-gongen-genki* scroll. In 1351, dwarf trees displayed on short poles were portrayed in the *Boki Ekotoba* scroll. Several other scrolls and paintings also included depictions of these kinds of trees. A close relationship between Japan`s Zen Buddhism and the potted trees began to shape bonsai reputation and esthetics. In this period, Chinese Chan Buddhist monks taught at Japan's monasteries. One of the monks` activities was to introduce political leaders to various arts of miniature landscapes as admirable accomplishments for men of taste and learning. Potted landscape arrangements up to this period included miniature figurines after the Chinese fashion. Japanese artists came to consider these items unnecessary, simplifying their creations in the spirit of Zen Buddhism.

Hachi-No-Ki

Chinese Penjing specimen with decorated and relatively deep ("bowl"-style) container

Around the 14th century, the term for dwarf potted trees was "the bowl`s tree". This indicated use of a fairly deep pot, rather than the shallow pot denoted by the eventual term *bonsai*. *Hachi no Ki* (*The Potted Trees*) is also the title of a Noh play by Zeami Motokiyo (1363–1444), based on a story c. 1383 about an impoverished samurai who burns his last three dwarf potted trees as firewood to warm a traveling monk. The monk is a disguised official who later rewards the samurai for his actions. In later centuries, woodblock prints by several artists depicted this popular drama. There was even a fabric design of the same name. Through these and other popular media, bonsai became known to a broad Japanese population.

Bonsai cultivation reached a high level of expertise in this period. Bonsai dating to the 17th century have survived to the present. One of the oldest-known living bonsai trees, considered one of the National Treasures of Japan, can be seen in the Tokyo Imperial Palace collection. A five-needle pine (*Pinus pentaphylla* var. *negishi*) known as Sandai-Shogun-No Matsu is documented as having been cared for by Tokugawa Iemitsu. The tree is thought to be at least 500 years old and was trained as a bonsai by, at latest, the year 1610.

By the end of the 18th century, bonsai cultivation in Japan was becoming widespread and began to interest the general public. In the Tenmei era (1781–88), an exhibit of traditional dwarf potted pines began to be held every year in Kyoto. Connoisseurs from five provinces and neighboring areas would bring one or two plants each to the show in order to submit them to visitors for ranking.

Classical Period

Depicting foliage in the Manual of the Mustard Seed Garden

In Japan after 1800, bonsai began to move from being the esoteric practice of a few specialists to becoming a widely popular art form and hobby. In Itami, Hyōgo (near Osaka),

Japanese scholars of Chinese arts gathered in the early 19th century to discuss recent styles in the art of miniature trees. Many terms and concepts adopted by this group were derived from *Kai-shi-en Gaden*, the Japanese version of *Jieziyuan Huazhuan* (Manual of the Mustard Seed Garden). The Japanese version of potted trees, which had been previously called "bunjin ueki", "bunjin hachiue", or other terms, were renamed "bonsai" (the Japanese pronunciation of the Chinese term penzai). This word connoted a shallow container, not a deeper bowl style. The term "bonsai", however, would not become broadly used in describing Japan`s dwarf potted trees for nearly a century.

The popularity of bonsai began to grow outside the limited scope of scholars and the nobility. On October 13, 1868, the Meiji Emperor moved to his new capital in Tokyo. Bonsai were displayed both inside and outside Meiji Palace, and those placed in the grand setting of the Imperial Palace had to be "Giant Bonsai," large enough to fill the grand space. The Meiji Emperor encouraged interest in bonsai, which broadened its importance and appeal to his government`s professional staff.

New books, magazines, and public exhibitions made bonsai more accessible to the Japanese populace. An Artistic Bonsai Concours was held in Tokyo in 1892, followed by publication of a three-volume commemorative picture book. This event demonstrated a new tendency to see bonsai as an independent art form. In 1903, the Tokyo association Jurakukai held showings of bonsai and ikebana at two Japanese-style restaurants. Three years later, *Bonsai Gaho* (1906 to c. 1913), became the first monthly magazine on the subject. It was followed by *Toyo Engei* and *Hana* in 1907. The initial issue of *Bonsai* magazine was published in 1921 by Norio Kobayashi (1889–1972), and this influential periodical would run for 518 consecutive issues.

Modern bonsai tools (left to right): leaf trimmer; rake with spatula; root hook; coir brush; concave cutter; knob cutter; wire cutter; small, medium and large shears

Bonsai shaping aesthetics, techniques, and tools became increasingly sophisticated as bonsai`s popularity grew in Japan. In 1910, shaping with wire rather than the older string, rope, and burlap techniques, appeared in the Sanyu-en Bonsai-Dan (History of Bonsai in the Sanyu nursery). Zinc-galvanized steel wire was initially used. Expensive copper wire was used only for selected trees that had real potential. In the 1920s and 1930s, Toolsmith Masakuni I (1880–1950) helped design and produce the first steel tools specifically made for the developing requirements of bonsai styling. These included the concave cutter, a branch cutter designed to leave a shallow indentation on the trunk when a branch was removed. Properly treated, this indentation would fill over with live tree tissue and bark over time, greatly reducing or eliminating the usual pruning scar.

Prior to World War II, international interest in bonsai was fueled by increased trade in trees and the appearance of books in popular foreign languages. By 1914, the first national annual bonsai show was held (an event repeated annually through 1933) in Tokyo`s Hibiya Park. Another great annual public exhibition of trees began in 1927 at the Asahi Newspaper Hall in Tokyo. Beginning in 1934, the prestigious Kokufu-ten annual exhibitions were held in Tokyo`s Ueno Park. The first major book on the subject in English was published in the Japanese capital: *Dwarf Trees (Bonsai)* by Shinobu Nozaki (1895–1968).

By 1940, about 300 bonsai dealers worked in Tokyo. Some 150 species of trees were being cultivated, and thousands of specimens annually were shipped to Europe and America. The first bonsai nurseries and clubs in the Americas were started by first and second-generation Japanese immigrants. Though this progress to international markets and enthusiasts was interrupted by the war, bonsai had by the 1940s become an art form of international interest and involvement.

Modern Bonsai

Following World War II, a number of trends made the Japanese tradition of bonsai increasingly accessible to Western and world audiences. One key trend was the increase in the number, scope, and prominence of bonsai exhibitions. For example, the Kokufu-ten bonsai displays reappeared in 1947 after a four-year cancellation and became annual affairs. These displays continue to this day, and are by invitation only for eight days in February. In October 1964, a great exhibition was held in Hibya Park by the private Kokufu Bonsai Association, reorganized into the Nippon Bonsai Association, to mark the Tokyo Olympics.

A large display of bonsai and suiseki was held as part of Expo '70, and formal discussion was made of an international association of enthusiasts. In 1975, the first Gafu-ten (Elegant-Style Exhibit) of *shohin* bonsai (13–25 cm (5–10 in) tall) was held. So was the first Sakufu-ten (Creative Bonsai Exhibit), the only event in which professional bonsai growers exhibit traditional trees under their own names rather than under the name of the owner.

The First World Bonsai Convention was held in Osaka during the World Bonsai and Suiseki Exhibition in 1980. Nine years later, the first World Bonsai Convention was held in Omiya and the World Bonsai Friendship Federation (WBFF) was inaugurated. These conventions attracted several hundreds of participants from dozens of countries and have since been held every four years at different locations around the globe: 1993, Orlando, Florida; 1997, Seoul, Korea; 2001, Munich, Germany; 2005, Washington, D.C.; 2009, San Juan, Puerto Rico. Currently, Japan continues to host regular exhibitions with the world`s largest numbers of bonsai specimens and the highest recognized specimen quality.

Another key trend was the increase in books on bonsai and related arts, now being published for the first time in English and other languages for audiences outside Japan.

In 1952, Yuji Yoshimura, son of a leader in the Japanese bonsai community, collaborated with German diplomat and author Alfred Koehn to give bonsai demonstrations. Koehn had been an enthusiast before the war, and his 1937 book *Japanese Tray Landscapes* had been published in English in Peking. Yoshimura`s 1957 book *The Art of Bonsai*, written in English with his student Giovanna M. Halford, went on to be called the "classic Japanese bonsai bible for westerners" with over thirty printings.

Multi-species saikei named *Roan Mountain* contains Shimpaku juniper and Zakura azalea.

The related art of saikei was introduced to English-speaking audiences in 1963 in Kawamoto and Kurihara`s book *Bonsai-Saikei*. This book described tray landscapes made with younger plant material than was traditionally used in bonsai, providing an alternative to the use of large, older plants, few of which had escaped war damage.

A third trend was the increasing availability of expert bonsai training, at first only in Japan and then more widely. In 1967 the first group of Westerners studied at an Ōmiya nursery. Returning to the U.S., these people established the American Bonsai Society. Other groups and individuals from outside Asia then visited and studied at the various Japanese nurseries, occasionally even apprenticing under the masters. These visitors brought back to their local clubs the latest techniques and styles, which were then further disseminated. Japanese teachers also traveled widely, bringing hands-on bonsai expertise to all six continents

The final trend supporting world involvement in bonsai is the widening availability of specialized bonsai plant stock, soil components, tools, pots, and other accessory items. Bonsai nurseries in Japan advertise and ship specimen bonsai worldwide. Most countries have local nurseries providing plant stock as well. Japanese bonsai soil components, such as Akadama clay, are available worldwide, and suppliers also provide similar local materials in many locations. Specialized bonsai tools are widely available from Japanese and Chinese sources. Potters around the globe provide material to hobbyists and specialists in many countries.

Bonsai has now reached a worldwide audience. There are over twelve hundred books on bonsai and the related arts in at least twenty-six languages available in over ninety countries and territories. A few dozen magazines in over thirteen languages are in print. Several score of club newsletters are available on-line, and there are at least that many discussion forums and blogs. There are at least a hundred thousand enthusiasts in some fifteen hundred clubs and associations worldwide, as well as over five million unassociated hobbyists. Plant material from every location is being trained into bonsai and displayed at local, regional, national, and international conventions and exhibitions for enthusiasts and the general public.

Cultivation and Care

Bonsai cultivation and care requires techniques and tools that are specialized to support the growth and long-term maintenance of trees in small containers.

Material Sources

All bonsai start with a specimen of source material, a plant that the grower wishes to train into bonsai form. Bonsai practice is an unusual form of plant cultivation in that growth from seeds is rarely used to obtain source material. To display the characteristic aged appearance of a bonsai within a reasonable time, the source plant is often mature or at least partially grown when the bonsai creator begins work. Sources of bonsai material include:

- *Propagation* from a source tree through cuttings or layering.
- *Nursery stock* directly from a nursery, or from a garden centre or similar resale establishment.
- *Commercial bonsai growers*, which, in general, sell mature specimens that display bonsai aesthetic qualities already.
- *Collecting* suitable bonsai material in its original wild situation, successfully moving it, and replanting it in a container for development as bonsai. These trees are called yamadori and are often the most expensive and prized of all Bonsai.

Techniques

This juniper makes extensive use of both jin (deadwood branches) and shari (trunk deadwood).

The practice of bonsai development incorporates a number of techniques either unique to bonsai or, if used in other forms of cultivation, applied in unusual ways that are particularly suitable to the bonsai domain. These techniques include:

- *Leaf trimming*, the selective removal of leaves (for most varieties of deciduous tree) or needles (for coniferous trees and some others) from a bonsai's trunk and branches.
- *Pruning* the trunk, branches, and roots of the candidate tree.
- *Wiring* branches and trunks allows the bonsai designer to create the desired

general form and make detailed branch and leaf placements.

- *Clamping* using mechanical devices for shaping trunks and branches.
- *Grafting* new growing material (typically a bud, branch, or root) into a prepared area on the trunk or under the bark of the tree.
- *Defoliation*, which can provide short-term dwarfing of foliage for certain deciduous species.
- *Deadwood bonsai techniques* such as *jin* and *shari* simulate age and maturity in a bonsai.

Care

Small trees grown in containers, like bonsai, require specialized care. Unlike houseplants and other subjects of container gardening, tree species in the wild, in general, grow roots up to several meters long and root structures encompassing several thousand liters of soil. In contrast, a typical bonsai container is under 25 centimeters in its largest dimension and 2 to 10 liters in volume. Branch and leaf (or needle) growth in trees is also of a larger scale in nature. Wild trees typically grow 5 meters or taller when mature, whereas the largest bonsai rarely exceed 1 meter and most specimens are significantly smaller. These size differences affect maturation, transpiration, nutrition, pest resistance, and many other aspects of tree biology. Maintaining the long-term health of a tree in a container requires some specialized care techniques:

- *Watering* must be regular and must relate to the bonsai species' requirement for dry, moist, or wet soil.

- *Repotting* must occur at intervals dictated by the vigor and age of each tree.
- *Tools* have been developed for the specialized requirements of maintaining bonsai.
- *Soil composition and fertilization* must be specialized to the needs of each bonsai tree, although bonsai soil is almost always a loose, fast-draining mix of components.
- *Location and overwintering* are species-dependent when the bonsai is kept outdoors as different species require different light conditions. It is important to note that few of the traditional bonsai species can survive inside a typical house, due to the usually dry indoor climate.

Aesthetics

Bonsai aesthetics are the aesthetic goals characterizing the Japanese tradition of growing an artistically shaped miniature tree in a container. Many Japanese cultural characteristics, in particular the influence of Zen Buddhism and the expression of Wabi-sabi, inform the bonsai tradition in Japan. Established art forms that share some aesthetic principles with bonsai include penjing and saikei. A number of other cultures around the globe have adopted the Japanese aesthetic approach to bonsai, and, while some variations have begun to appear, most hew closely to the rules and design philosophies of the Japanese tradition.

Over centuries of practice, the Japanese bonsai aesthetic has encoded some important techniques and design guidelines. Like the aesthetic rules that govern, for example, Western common practice period music, bonsai`s guidelines help practitioners work within an established tradition with some assurance of success. Simply following the guidelines alone will not guarantee a successful result. Nevertheless, these design rules can rarely be broken without reducing the impact of the bonsai specimen. Some key principles in bonsai aesthetics include:

- *Miniaturization*: By definition, a bonsai is a tree kept small enough to be container-grown while otherwise fostered to have a mature appearance.
- *Proportion among elements*: The most prized proportions mimic those of a full-grown tree as closely as possible. Small trees with large leaves or needles are out of proportion and are avoided, as is a thin trunk with thick branches.
- *Asymmetry*: Bonsai aesthetics discourage strict radial or bilateral symmetry in branch and root placement.
- *No trace of the artist*: The designer's touch must not be apparent to the viewer. If a branch is removed in shaping the tree, the scar will be concealed. Likewise, wiring should be removed or at least concealed when the bonsai is shown, and must leave no permanent marks on the branch or bark.
- *Poignancy*: Many of the formal rules of bonsai help the grower create a tree that expresses Wabi-sabi, or portrays an aspect of *mono no aware*.

Display

Bonsai displayed on an outdoor bench - note automated watering apparatus

A Seiju elm bonsai on display with a shitakusa of miniature hosta and a hanging scroll.

A bonsai display presents one or more bonsai specimens in a way that allows a viewer to see all the important features of the bonsai from the most advantageous position. That position emphasizes the bonsai`s defined "front", which is designed into all bonsai. It places the bonsai at a height that allows the viewer to imagine the bonsai as a full-size tree seen from a distance, siting the bonsai neither so low that the viewer appears to be hovering in the sky above it nor so high that the viewer appears to be looking up at the tree from beneath the ground. Noted bonsai writer Peter Adams recommends that bonsai be shown as if "in an art gallery: at the right height; in isolation; against a plain background, devoid of all redundancies such as labels and vulgar little accessories."

For outdoor displays, there are few aesthetic rules. Many outdoor displays are semi-permanent, with the bonsai trees in place for weeks or months at a time. To avoid damaging the trees, therefore, an outdoor display must not impede the amount of sunlight needed for the trees on display, must support watering, and may also have to block excessive wind or precipitation. As a result of these practical constraints, outdoor displays are often rustic in style, with simple wood or stone components. A common design is the bench, sometimes with sections at different heights to suit different sizes of bonsai, along which bonsai are placed in a line. Where space allows, outdoor bonsai specimens are spaced far enough apart that the viewer can concentrate on one at a time. When the trees are too close to each other, aesthetic discord between adjacent trees of different sizes or styles can confuse the viewer, a problem addressed by exhibition displays.

Exhibition displays allow many bonsai to be displayed in a temporary exhibition format, typically indoors, as would be seen in a bonsai design competition. To allow many trees to be located close together, exhibition displays often use a sequence of small alcoves, each containing one pot and its bonsai contents. The walls or dividers between the alcoves make it easier to view only one bonsai at a time. The back of the alcove is a neutral color and pattern to avoid distracting the viewer`s eye. The bonsai pot is almost always placed on a formal stand, of a size and design selected to complement the bonsai and its pot.

Indoors, a formal bonsai display is arranged to represent a landscape, and traditionally consists of the featured bonsai tree in an appropriate pot atop a wooden stand, along with a shitakusa (companion plant) representing the foreground, and a hanging scroll representing the background. These three elements are chosen to complement each other and evoke a particular season, and are composed asymmetrically to mimic nature. When displayed inside a traditional Japanese home, a formal bonsai display will often be placed within the home's tokonoma or formal display alcove. An indoor display is usually very temporary, lasting a day or two, as most bonsai are intolerant of indoor conditions and lose vigor rapidly within the house.

Containers

A variety of informal containers may house the bonsai during its development, and even trees that have been formally planted in a bonsai pot may be returned to growing boxes from time to time. A large growing box can house several bonsai and provide a great volume of soil per tree to encourage root growth. A training box will have a single specimen, and a smaller volume of soil that helps condition the bonsai to the eventual size and shape of the formal bonsai container. There are no aesthetic guidelines for these development containers, and they may be of any material, size, and shape that suit the grower.

Assorted bonsai pots

Completed trees are grown in formal bonsai containers. These containers are usually ceramic pots, which come in a variety of shapes and colors and may be glazed or unglazed. Unlike many common plant containers, bonsai pots have drainage holes in the bottom surface to complement fast-draining bonsai soil, allowing excess water to escape the pot. Growers cover the holes with a screening to prevent soil from falling out and to hinder pests from entering the pots from below. Pots usually have vertical sides, so that the tree's root mass can easily be removed for inspection, pruning, and replanting, although this is a practical consideration and other container shapes are acceptable.

There are alternatives to the conventional ceramic pot. Multi-tree bonsai may be created atop a fairly flat slab of rock, with the soil mounded above the rock surface and the trees planted within the raised soil. In recent times, bonsai creators have also begun to fabricate rock-like slabs from raw materials including concrete and glass-reinforced plastic. Such constructed surfaces can be made much lighter than solid rock, can include depressions or pockets for additional soil, and can be designed for drainage of water, all characteristics difficult to achieve with solid rock slabs. Other unconventional containers can also be used, but in formal bonsai display and competitions in Japan, the ceramic bonsai pot is the most common container.

For bonsai being shown formally in their completed state, pot shape, color, and size are chosen to complement the tree as a picture frame is chosen to complement a painting. In general, containers with straight sides and sharp corners are used for formally shaped plants, while oval or round containers are used for plants with informal designs. Many aesthetic guidelines affect the selection of pot finish and color. For example, evergreen bonsai are often placed in unglazed pots, while deciduous trees usually appear in glazed pots. Pots are also distinguished by their size. The overall design of the bonsai tree, the thickness of its trunk, and its height are considered when determining the size of a suitable pot.

Some pots are highly collectible, like ancient Chinese or Japanese pots made in regions with experienced pot makers such as Tokoname, Japan or Yixing, China. Today many potters worldwide produce pots for bonsai.

Bonsai Styles

Formal upright style Bald cypress

Informal upright style Juniper

Slant-style conifer

Cascade style conifer

Forest style Black Hills Spruce

The Japanese tradition describes bonsai tree designs using a set of commonly understood, named styles.The most common styles include formal upright, informal upright, slanting, semi-cascade, cascade, raft, literati, and group/forest. Less common forms include windswept, weeping, split-trunk, and driftwood styles.These terms are not mutually exclusive, and a single bonsai specimen can exhibit more than one style characteristic. When a bonsai specimen falls into multiple style categories, the common practice is to describe it by the dominant or most striking characteristic.

A frequently used set of styles describes the orientation of the bonsai tree's main trunk. Different terms are used for a tree with its apex directly over the center of the trunk's entry into the soil, slightly to the side of that center, deeply inclined to one side, and inclined below the point at which the trunk of the bonsai enters the soil.

- **Formal upright** or *chokkan* style trees are characterized by a straight, upright, tapering trunk. Branches progress regularly from the thickest and broadest at the bottom to the finest and shortest at the top.
- **Informal upright** or *moyogi* trees incorporate visible curves in trunk and branches, but the apex of the informal upright is located directly above the trunk's entry into the soil line.
- **Slant-style** or *shakan* bonsai possess straight trunks like those of bonsai grown in the formal upright style. However, the slant style trunk emerges from the soil at an angle, and the apex of the bonsai will be located to the left or right of the root base.
- **Cascade-style** or *kengai* specimens are modeled after trees that grow over water or down the sides of mountains. The apex (tip of the tree) in the semi-cascade-style or *han kengai* bonsai extend just at or beneath the lip of the bonsai pot; the apex of a (full) cascade style falls below the base of the pot.

A number of styles describe the trunk shape and bark finish. For example, the deadwood bonsai styles identify trees with prominent dead branches or trunk scarring.

- **Shari** or *sharimiki* style involves portraying a tree in its struggle to live while a significant part of its trunk is bare of bark.

Although most bonsai trees are planted directly into the soil, there are styles describing trees planted on rock.

- **Root-over-rock** or *sekijoju* is a style in which the roots of the tree are wrapped around a rock, entering the soil at the base of the rock.
- **Growing-in-a-rock** or *ishizuke* or *ishitsuki* style means the roots of the tree are growing in soil contained within the cracks and holes of the rock.

While the majority of bonsai specimens feature a single tree, there are well-established style categories for specimens with multiple trunks.

- **Forest (or group)** or *yose ue* style comprises a planting of several or many trees of one species, typically an odd number, in a bonsai pot.
- **Multi-trunk** styles like *sokan* and *sankan* have all the trunks growing out of one spot with one root system, so the bonsai is actually a single tree.
- **Raft-style** or *ikadabuki* bonsai mimic a natural phenomenon that occurs when a tree topples onto its side, for example, from erosion or another natural force. Branches along the top side of the trunk continue to grow as a group of new trunks.

Other Styles

A few styles do not fit into the preceding categories. These include:

- **Literati** or *bunjin-gi* style is characterized by a generally bare trunk line, with branches reduced to a minimum, and foliage placed toward the top of a long, often contorted trunk.
- **Broom** or *hokidachi* style is employed for trees with fine branching, like elms. The trunk is straight and branches out in all directions about ⅓ of the way up the entire height of the tree. The branches and leaves form a ball-shaped crown.
- **Windswept** or *fukinagashi* style describes a tree that appears to be affected by strong winds blowing continuously from one direction, as might shape a tree atop a mountain ridge or on an exposed shoreline.

Size Classifications

Japanese bonsai exhibitions and catalogs frequently refer to the size of individual bonsai specimens by assigning them to size classes (see table below). Not all sources agree on the exact sizes or names for these size ranges, but the concept of the ranges is well-established and useful to both the cultivation and the aesthetic understanding of the trees. A photograph of a bonsai may not give the viewer an accurate impression of the tree's real size, so printed documents may complement a photograph by naming the bonsai's size class. The size class implies the height and weight of the tree in its container.

In the very largest size ranges, a recognized Japanese practice is to name the trees "two-handed", "four-handed", and so on, based on the number of men required to move the tree and pot. These trees will have dozens of branches and can closely simulate a full-size tree. The very largest size, called "imperial", is named after the enormous potted trees of Japan's Imperial Palace.

At the other end of the size spectrum, there are a number of specific techniques and styles associated solely with the smallest common sizes, mame and shito. These techniques take advantage of the bonsai's minute dimensions and compensate for the limited number of branches and leaves that can appear on a tree this small.

Common names for bonsai size classes		
Large bonsai		
Common name	**Size class**	**Tree Height**
Imperial bonsai	Eight-handed	152–203 cm (60–80 in)
Hachi-uye	Six-handed	102–152 cm (40–60 in)
Dai	Four-handed	76–122 cm (30–48 in)
Omono	Four-handed	76–122 cm (30–48 in)
Medium-size bonsai		
Common name	**Size class**	**Tree Height**
Chiu	Two-handed	41–91 cm (16–36 in)
Chumono	Two-handed	41–91 cm (16–36 in)
Katade-mochi	One-handed	25–46 cm (10–18 in)
Miniature bonsai		

Common name	Size class	Tree Height
Komono	One-handed	15–25 cm (6–10 in)
Shohin	One-handed	13–20 cm (5–8 in)
Mame	Palm size	5–15 cm (2–6 in)
Shito	Fingertip size	5–10 cm (2–4 in)
Keshitsubo	Poppy-seed size	3–8 cm (1–3 in)

Indoor Bonsai

The Japanese tradition of bonsai does not include indoor bonsai, and bonsai appearing at Japanese exhibitions or in catalogs have been grown outdoors for their entire lives. In less-traditional settings, including climates more severe than Japan's, indoor bonsai may appear in the form of potted trees cultivated for the indoor environment.

Traditionally, bonsai are temperate climate trees grown outdoors in containers. Kept in the artificial environment of a home, these trees weaken and die. But a number of tropical and sub-tropical tree species will survive and grow indoors. Some of these tropical and sub-tropical species are suited to bonsai aesthetics and can be shaped much as traditional outdoor bonsai are.

TREE FORK

A tree fork is a bifurcation in the trunk of a tree giving rise to two roughly equal diameter branches. These forks are a common feature of tree crowns. The wood grain orientation at the top of a tree fork is such that wood cells interlock to provide sufficient mechanical support. A common malformation of a tree fork is where bark has formed within the join and these bark-included junctions often have a heightened risk of failure.

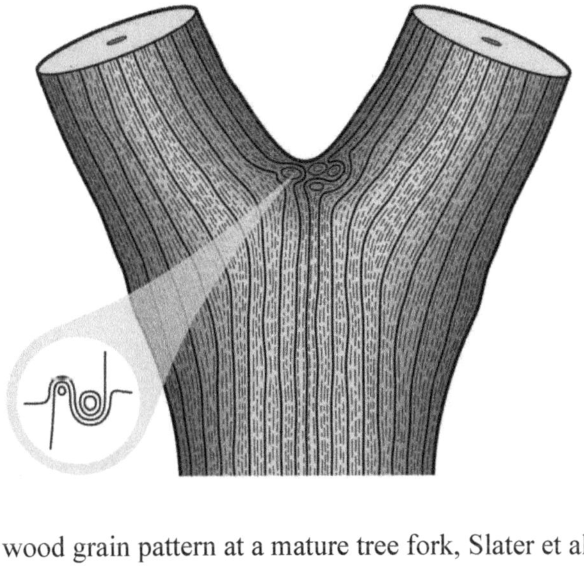

Typical wood grain pattern at a mature tree fork, Slater et al.2014

A typical tree fork in a Norway maple (Acer platanoides)

Wood grain pattern of a fork of Fraxinus excelsior (Common ash), as found by stripping off the outer and inner bark layers

Definition of a Tree Fork

In arboriculture, junctions in the crown structure of trees are frequently categorised as either branch-to-stem attachments or co-dominant stems. Co-dominant stems are where the two or more arising branches emerging from the junction are of near equal diameter and this type of junction in a tree is often referred to in layman's terms as 'a tree fork'.

There is actually no hard botanical division between these two forms of tree junction: they are topologically equivalent, and from their external appearance it is only a matter of the diameter ratio between the branches that are conjoined that separates a tree fork from being a branch-to-stem junction. However, when a small branch joins to a tree trunk there is a knot that can be found to be embedded into the trunk of the tree, which was the initial base of the smaller branch. This is not the case in tree forks, as each branch is roughly equal in size and no substantial tissues from either branch is embedded into the other, so there is no reinforcing knot to supply the mechanical strength to the junction that will be needed to hold the branches aloft.

Tree Fork Anatomy and Morphology

Research has shown that a unique wood grain pattern at the apex of forks in hazel trees (*Corylus avellana* L.) acts to hold together the branches in this species, and this is probably the case in most other woody plants and trees. This is an example of 'trade-off' in xylem, where mechanical strength to the tree`s junction is gained at the expense of efficiency in tree sap conductance The complex interlocking wood grain patterns developed within tree forks present a great opportunity for biomimicry (the mimicking of natural sbiological structures in man-made materials) in fibrous materials, where the production of a Y-shaped or T-shaped component is needed: particularly in such components that may need to act as a conduit for liquids as well as being mechanically strong.

Tree fork morphology has been shown to alter with the angle of inclination of the fork from the vertical axis. As the fork becomes more tilted from the vertical, the branches become more elliptical in cross-section, to adapt to the additional lateral loading upon them, and Buckley, Slater and Ennos (2015) showed that this adaption resulted in stronger tree forks when the fork was more inclined away from the vertical.

Bark Inclusions and Fork Strength

Where a junction forms in a tree and bark is incorporated into the join, this is referred to as an 'included bark junction' or 'bark inclusion'. These can be substantially weaker in strength than normal tree forks, and can become a significant hazard in a tree. Research has shown that in hazel trees, the more the included bark is occluded within new wood growth, the stronger that junction will be, with the weakest forks being those with a large amount of unoccluded bark at their apex. Common tree care practices are to prune out such bark-included forks at an early stage of the tree's development, to brace the two arising branches above such a junction so that they can not split apart (using a flexible brace) or to reduce the length of the smaller arising branch, so that it is subordinated to the larger branch.

An included bark junction formed in a wild cherry tree (*Prunus avium*)

The strength of a normally-formed tree fork can be assessed by its shape: those that are more U-shaped are typically considerably stronger than those that are V-shaped at their apex. This characteristic, and the presence of bark included in a tree fork, are important attributes for tree surveyors and tree contractors to note in order to assess whether the tree fork is a defect in the structure of a tree.

REFERENCES

1. Stebbins, R.L. (November 2007). "Training & Pruning Your Home Orchard (OSU)" (PDF). Pacific Northwest Extension publications Fionnuala Fallon (3 March 2012). "The trees that shape our lives". The Irish Times. Ireland.
2. Mörður Gunnarsson (2012). "Living Furniture". Cottage and Garden. Iceland. pp. 28–29.
3. Mark Primack. "Pleaching". The NSW Good Wood Guide. Retrieved 2010-05-10.
4. Lewin, Brent (2012), "November Volume 2012 Article", India's living Bridges, Reader's Digest Australia, pp. 82–89, EAN 9311484018704, archived from the original on 6 June 2013
5. Baker, Russ.""",| Business Insider, 6 Oct. 2011
6. "Living Growing Root Bridges Are 100% Natural Architecture". Retrieved 2013-04-12.
7. "Living Root Bridge". Online Highways LLC. 2005-10-21. Retrieved 2010-05-07.
8. "Cherrapunjee.com: A Dream Place". Cherrapunjee Holiday Resort. Retrieved 2010-05-07.

9. David Davies (1 June 1996). "Plant your own furniture. Watch it grow". The Independent. UK. Retrieved 15 August 2011.
10. Fischbacher, Thomas (2007), Botanical Engineering (PDF), School of Engineering Sciences @ University of Southampton, archived from the original(PDF) on 22 December 2009
11. "A very special tree house". Bio-pro.de. 2010-02-04. Retrieved 2010-04-14.
12. McKee, Kate (2012), "Living sculpture", Sustainable and water wise gardens, Westview: Universal Wellbeing PTY Limited, pp. 70–73
13. Richard Reames (2005), Arborsculpture: Solutions for a Small Planet, Oregon: Arborsmith Studios, ISBN 0-9647280-8-7
14. "Eco-Architecture Could Produce "Grow Your Own" Homes". American Friends of Tel Aviv University.
15. Erlandson, Wilma (2001), My father "talked to trees", Westview: Boulder, p. 22, ISBN 0-9708932-0-5
16. Volz, Martin (Oct–Nov 2008), "A Tree shaper's life." (PDF), Queensland Smart Farmer
17. j Richard Reames; Delbol, Barbara (1995), How to Grow a Chair: The Art of Tree Trunk Topiary, ISBN 0-9647280-0-1
18. "Treenovations". Treenovations. Treenovations. Retrieved 6 August 2015.
19. Carter, W.A. (1942). A method of growing plants in water vapor to facilitate examination of roots. Phytopathology 732: 623-625.
20. Stoner, R.J. (1983). Rooting in Air. Greenhouse Grower Vol I No. 11
21. US "A method of shaping a portion of a woody plant into a desired form is provided. The method is effected by providing a root of a woody plant, shaping the root into the desired form and culturing the root under conditions suitable for secondary thickening of the root." 7328532, Golan, Ezekiel, "Method and a kit for shaping a portion of a woody plant into a desired form", issued 2008-02-12
22. Swati Balgi (September 2009), "Live Art" (PDF), Society Interiors Magazine, Prabhadevi, Mumbai: Magna Publishing
23. chaika, Chaika (2013), Growing... furniture, Bulgarian
24. Rodkin, Dennis (25 February 1996), The Gardener, Chicago Tribune Sunday
25. Oommen, Ansel (15 September 2013), The Artful Science of Tree Shaping, www.permaculture.co.uk
26. Link, Tracey (13 June 2008), "Senior project for Bachelor of Science degree in Landscape Architecture", Arborsculpture: An Emerging Art Form and Solutions to our Environment (PDF), p. 41
27. Roger, Fox (December 2012), "Artist tree", Better Homes and Gardens Last, p. 140
28. Cattle, Christopher. "How to grow your stool". Retrieved 2010-06-14.
29. "Living Trees, Living Art". Retrieved 2009-05-08.

30. Peter Cook & Becky Northey (2012). Knowledge to Grow Shaped Trees. 1 (first ed.). Australia: SharBrin Publishing Ptd Ltd. ISBN 978-1-921571-54-1.
31. (PDF), Farm Show Magazine, vol. 32 no. 4, p. 9, June–August 2008, archived from the original (PDF) on 8 March 2012, retrieved 2010-05-08
32. Weston, Sarah (3 October 2006), Axel Erlandson's Tree Circus, Mid-County Post
33. Hicks, Ivan; Rosenfeld, Richard; Whitworth, Jo (2007), Tricks with Trees, Pavilion Books, p. 160, ISBN 1-86205-734-6
34. "home & your garden", Going on a 'bender', Queensland Times, May 2012, p. 18
35. Musitelui, Robin (30 May 1995), "Santa Cruz Sentinel Article", Architect rescue famous tree, p. 222
36. Ken Mudge; Jules Janick; Steven Scofield; Eliezer E. Goldschmidt (2009), Jules Janick, ed., A History of Grafting (PDF), Issues in New Crops and New Uses, Purdue University Center for New Crops and Plants Products, orig. pub. John Wiley & Sons, Inc., pp. 442–443Note large file: 8.04MB
37. Jim Worrall (27 May 2007), Forest and Shade Tree Pathology: Wood Decay, retrieved 2011-06-10
38. "BOTANY BUILDINGS Grow Buildings From Trees!". Retrieved 2013-04-12.
39. Menges, Achim (2 March 2012). Material Computation 'Higher Integration in Morphogenetic Design Architectural Design (Architectural Design)'. United Kingdom: John Wiley & Sons Ltd. p. 144. ISBN 0-470-97330-7.
40. Walpole, Lois (2004), grown home, retrieved 2010-06-14
41. University of California, Cooperative Extension (November 2003), "Arborsculpture: Horticultural Art"(PDF), Landscape & Turf News, p. 6, archived from the original (PDF) on 2010-06-10, retrieved 2015-12-12
42. Cassidy, Patti (August 2008), "A Truly Living Art", Rhode Island Home, Living and Design Magazine, Swansea, Massachusetts: Home, Living & Design, Inc., pp. 26–27
43. Reddy, Jini (2010-01-23). "Trail Of The Unexpected: The root masters of India". Cherrapunjee Holiday Resort. Retrieved 2010-05-08.
44. Mack, Daniel (1996-12-31), Making Rustic Furniture: The Tradition, Spirit, and Technique with Dozens of Project Ideas (illustrated ed.), Lark Books, p. 160, ISBN 1-887374-12-4
45. "Only Natural Grown Chair". Shawano Leader Newspaper. Wisconsin Historical Society. 1922-10-19. Retrieved 2010-05-15.
46. "Obituary of Axel Erlandson", Turlock Journal, p. 15, 30 April 1964
47. Wiechula, Arthur (1926) [1926], Wachsende Häuser aus lebenden Bäumen entstehend (Developing Houses from Living Trees), Verl. Naturbau-Ges, p. 320
48. "designboom:history of arborsculpture".
49. Ladd, Dan (2009-01-22), Sculpturefest 2008: Daniel Ladd, retrieved 2010-06-14
50. "Dan Ladd's home page". Dan Ladd. Retrieved 2010-05-09.

51. Extreme Nature: The Sculptures of Dan Ladd at Putney Library 10 October 2006.
52. Shaw, Kurt (11 August 2002), "Persephone Project promotes gardening as contemporary art medium", TribLiveNews, retrieved 2010-06-30
53. Steve, Rhodes (April 6, 2003), "No need to pull up a stump: Short of garden furniture?", Sunday Mail
54. "The father of Living Furniture", Bangkok Post, 16 January 1996
55. "Pooktre". Northey, Becky. Retrieved 2010-05-05.(Archived by WebCite at http://www.webcitation.org/5pVaujskD)
56. "Pooktre", Bricks & Mortar Magazine, 2008
57. "Pooktre". Northey, Becky. Retrieved 2010-10-18.
58. McKie, Fred (20 April 2005), "Warwick artist grows wooden 'jewels' for World Expo", The Southern Free Times
59. Tibballs, Geoff; Proud, James (2009), Ripley's Believe It or Not: Seeing is Believing, :Orlando, FL: Ripley Publishing Ltd, p. 32, ISBN 978-1-893951-45-7
60. Biography of Richard Reames, retrieved 2010-06-27
61. Company profile: Arborsmith Studios
62. S. Okenga (2001), Eden on Their Minds: American Gardeners with Bold Visions, Clarkson Potter, p. 110, ISBN 0-609-60587-9
63. Nestor, James (February 2007), "Branching Out", Dwell, Dwell, LLC, p. 96, retrieved 2010-06-15
64. Arbor Sculpture: "If you like I'll grow you a mirror" (PDF), The Cutting Edge; the Newsletter of the Victorian Woodworkers Association, Inc., June 2006, p. 16, archived from the original (PDF) on 25 October 2009, retrieved 2010-05-15
65. Life and Limbs
66. Cattle, Christopher. "grown furniture home page". Christopher Cattle. Retrieved 2010-06-14.
67. "Grown Furniture at the Museum of English Rural Life" (Press release). University of Reading, UK. 26 March 2008. Retrieved 2010-06-14.
68. Cattle, Christopher. "grown furniture examples". Christopher Cattle. Retrieved 2010-06-14.
69. "Five year deliveries", China Morning Business View, Farmington, Michigan: AccessMyLibrary, via CMP Information Ltd., via The Gale Group, 2003, retrieved 2010-06-15
70. Treet Them Well, Chaotic Web Development, via ananova.com), 2 February 2005, retrieved 2010-06-15
71. Hoffman, Bill; Wire Services (2005-02-03), "Weird But True", New York Post (news ed.), p. 23, retrieved 2010-06-15
72. Chan, Peter (1987), Bonsai Masterclass, Sterling Publishing Co., Inc., ISBN 0-8069-6763-3
73. Koreshoff, Deborah R. (1984), Bonsai: Its Art, Science, History and Philosophy,

Timber Press, Inc., p. 1, ISBN 0-88192-389-3

74. Evans, Erv, Espalier, North Carolina State University Horticultural Science Department Cooperative Extension Service, retrieved 2010-06-29
75. The Complete Guide to Pruning and Training Plants, Joyce and Brickell, 1992, page 106, Simon and Schuster
76. John Seymour (1984). The Forgotten Arts A practical guide to traditional skills. Angus & Robertson Publishers. p. 54. ISBN 0-207-15007-9.
77. Coombs, Duncan; Blackburne-Maze, Peter; Cracknell, Martyn; Bentley, Roger (2001), "9", The Complete Book of Pruning (illustrated ed.), Sterling Publishing Company, p. 224, ISBN 978-1-84188-143-0
78. "A LIVING HOUSE - Terreform's Fab Tree Hab". Retrieved 2013-04-14.
79. "Grow your own home: 'Fab tree hab'". Retrieved 2013-04-14.
80. Karen Cilento (28 October 2011), "The Patient Gardener / Visiondivision", Arch Daily, Plataforma Networks, retrieved 8 March 2012
81. Foer, Joshua; Reames, Richard (Winter 2005–2006), "How to Grow a Chair: An Interview with Richard Reames", Cabinet Magazine, retrieved 2010-05-15.
82. Jules Janick (2009), Horticultural Reviews, 35, John Wiley and Sons, p. 443, ISBN 978-0-470-38642-2
83. Jaya Jiwatram (2008-08-25), "We're going to Live in the Trees", Popular Science Magazine, retrieved 2011-06-10
84. Hao Jinyao (11 May 2009), "The art of Tree shaping", Culture
85. Marras, Amerigo (1 February 1999), ECO-TEC: Architecture of the In-Between, Princeton Architectural Press, ISBN 978-1-56898-159-8
86. Stephen Lesiuk. "BIOTECTURE II: PLANTBUILDING INTERACTION". Dept of Architecture, Sydney University. Retrieved 2011. Check date values in: |access-date= (help)
87. "Living Art". Discoveries. 6 September 2011. Archived from the original on 5 December 2010.[dead link]
88. Varkulevicius, Jane (2010), Pruning for Flowers and Fruit, CSIRO Publishing, p. 96
89. Bunny Guinness (18 Sep 2011), "Train your trees into extraordinary shapes", Sunday Telegraph, UK/
90. "designboom: the alchemic force of the imagination transmutes nature". Retrieved 2011-06-10.
91. Perréal, Jean (1516). "l'Alchimie". Musée Marmottan Monet. Archived from the original on 19 March 2009. Retrieved 2010-05-08.
92. Kamil, Neil (2005), Fortress of the Soul: Violence, Metaphysics, and Material Life in the Huguenots' New World 1517–1751, JHU Press, pp. 384–385, ISBN 0-8018-7390-8, retrieved 2010-02-22
93. Swedenborg, Emanuel (2008) [1758], Earths in the Universe, BiblioBazaar,

LLC, p. 104, ISBN 1-4375-3106-7

94. Lorber, Jakob (1995), Die Haushaltung Gottes (The Household of God) (Translation by Violet Ozols ed.), Lorber Verlag, p. 564, ISBN 978-3-87495-314-6
95. Slater, D. Bradley, R, Withers, P. and Ennos, A. R. (2014) The anatomy and grain pattern in forks of hazel (Corylus avellana L.) and other tree species, Trees 28 (5) 1437-1448
96. Smiley, E. T. (2003) Does included bark reduce the strength of co-dominant stems?; Journal of Arboriculture 29(2) pp. 104-106
97. Harris, R.W. Clark, J. R. and Matheny, N. P. (1999) Arboriculture: Integrated Management of Landscape Trees, Shrubs and Vines; 3rd Ed.; Prentice Hall
98. Lonsdale, D.(2000) Hazards from Trees: A general guide; Forestry Commission Practice Guide; TSO
99. Slater, D. and Harbinson, C. J. (2010) Towards a new model for branch attachment; Arboricultural Journal 33 (2) 95-105
100. Slater, D. and Harbinson, C. J. (2010) Towards a new model for branch attachment; Arboricultural Journal 33 (2) 95-105
101. Slater, D. and Ennos, A. R. (2013) Determining the mechanical properties of hazel forks by testing their component parts, Trees 27 (6) 1515-1524
102. Slater, D. Bradley, R, Withers, P. and Ennos, A. R. (2014) The anatomy and grain pattern in forks of hazel (Corylus avellana L.) and other tree species, Trees 28 (5) 1437-1448
103. Slater, D. and Ennos, A. R. (2015) Interlocking wood grain patterns provide improved wood strength in forks of hazel (Corylus avellana L.), Arboricultural Journal 37 (1) 1-12
104. Domec J.C. & Gartner B.L. (2001) Cavitation and water storage capacity in bole xylem segments of mature and young Douglas-fir trees. Trees 15, 204–214
105. Burns L.A., Mouritz A.P.,Pook D. and Feih, S. (2012) Bio-inspired design of aerospace composite joints for improved damage tolerance; Composite Structures 94 995–1004
106. Buckley, G. Slater, D. and Ennos, A. R. (2015) Angle of inclination affects the morphology and strength of bifurcations in hazel (Corylus avellana L.), Arboricultural Journal 37 (2) 99-112
107. Lonsdale, D. (1999) Principles of Tree Hazard Assessment and Management, TSO: London
108. Smiley, E. T. (2003) Does included bark reduce the strength of co-dominant stems? Journal of Arboriculture 29 (2): 104-106
109. Slater, D. and Ennos, A. R. (2015) The level of occlusion of included bark affects the strength of bifurcations in hazel (Corylus avellana L.) Journal of Arboriculture and Urban Forestry 41 (4): 194-207
110. Gilman E. F. (2012) Illustrated guide to pruning; third edition; Delmar publishers, Albany, NY.

111. Turner, S., Slater, D. and Ennos, A. R. (2012) Failure of forks in clonal varieties of Platanus x acerifolia; Arboricultural Journal 34 179-189

112. Lonsdale, D.(2000) Hazards from Trees: A general guide; Forestry Commission Practice Guide; TSO

CHAPTER 4

Landscape Architecture

Chapter Contents

- **Introduction**
- **Definition**
- **Fields of Activity**
- **History of Landscape Architecture**
- **Relation to Urban Planning**
- **Profession**
- **Examples of Landscape Architecture**

Landscape architecture is a branch of architecture that deals with the planning and design of land and its relation to the buildings around it. This practice blends site planning, landscaping, art, and environmental restoration to help connect an area to the buildings around it, and make the landscape attractive in itself. The goal of landscape architecture is to create pleasing, functional, and beautiful spaces that serve the needs of their owners or the public.

INTRODUCTION

Landscape architecture is the design of outdoor public areas, landmarks, and structures to achieve environmental, social-behavioral, or aesthetic outcomes. It involves the systematic investigation of existing social, ecological, and soil conditions and processes in the landscape, and the design of interventions that will produce the desired outcome. The scope of the profession includes landscape design; site planning; stormwater management; environmental restoration; parks and recreation planning; visual resource management; green infrastructure planning and provision; and private estate and residence landscape master planning and design; all at varying scales of design, planning and management. A practitioner in the profession of landscape architecture is called a landscape architect.

Stourhead in Wiltshire, England, designed by Henry Hoare (1705–1785), "the first landscape gardener, who showed in a single work, genius of the highest orders"

DEFINITION

Landscape architecture is a multi-disciplinary field, incorporating aspects of botany, horticulture, the fine arts, architecture, industrial design, soil sciences, environmental psychology, geography, and ecology. The activities of a landscape architect can range from the creation of public parks and parkways to site planning for campuses and corporate office parks, from the design of residential estates to the design of civil infrastructure and the management of large wilderness areas or reclamation of degraded landscapes such as mines or landfills. Landscape architects work on structures and external spaces with limitations toward the landscape or park aspect of the design - large or small, urban, suburban and rural, and with "hard" (built) and "soft" (planted) materials, while integrating ecological sustainability. The most valuable contribution can be made at the first stage of a project to generate ideas with technical understanding and creative flair for the design, organization, and use of spaces. The landscape architect can conceive the overall concept and prepare the master plan, from which detailed design drawings and technical specifications are prepared. They can also review proposals to authorize and supervise contracts for the construction work. Other skills include preparing design impact assessments, conducting environmental assessments and audits, and serving as an expert witness at inquiries on land use issues.

FIELDS OF ACTIVITY

Royal Botanic Gardens, Kew, London, established 1759The Palm House built 1844–1848 by Richard Turner to Decimus Burton's designs

Urban design in city squares. Water feature in London, by Tadao Ando who also works with landscapes and gardens

The variety of the professional tasks that landscape architects collaborate on is very broad, but some examples of project types include:

- The planning, form, scale and siting of new developments
- Parks of General design and public infrastructure
- Sustainable development
- Stormwater management including rain gardens, green roofs, groundwater recharge, Green infrastructure, and constructed wetlands
- Landscape design for educational function and site design for public institutions and government facilities
- Parks, botanical gardens, arboretums, greenways, and nature preserves
- Recreation facilities; i.e.: playgrounds, golf courses, theme parks and sports

facilities

- Housing areas, industrial parks and commercial developments
- Estate and residence landscape master planning and design
- Highways, transportation structures, bridges, and transit corridors
- Urban design, town and city squares, waterfronts, pedestrian schemes, and parking lots
- Natural park, tourist destination, and recreating historical landscapes, and historic garden appraisal and conservation studies
- Reservoirs, dams, power stations, reclamation of extractive industry applications or major industrial projects and mitigation
- Environmental assessment and landscape assessment, planning advice and land management proposals.
- Coastal and offshore developments and mitigation
- Ecological Design any aspect of design that minimizes environmentally destructive impacts by integrating itself with natural processes and sustainability

Landscape managers use their knowledge of landscape processes to advise on the long-term care and development of the landscape. They often work in forestry, nature conservation and agriculture.

Landscape scientists have specialist skills such as soil science, hydrology, geomorphology or botany that they relate to the practical problems of landscape work. Their projects can range from site surveys to the ecological assessment of broad areas for planning or management purposes. They may also report on the impact of development or the importance of particular species in a given area.

Landscape planners are concerned with landscape planning for the location, scenic, ecological and recreational aspects of urban, rural and coastal land use. Their work is embodied in written statements of policy and strategy, and their remit includes master planning for new developments, landscape evaluations and assessments, and preparing countryside management or policy plans. Some may also apply an additional specialism such as landscape archaeology or law to the process of landscape planning.

Green roof (or more specifically, vegetative roof) designers design extensive and intensive roof gardens for storm water management, evapo-transpirative cooling, sustainable architecture, aesthetics, and habitat creation.

HISTORY OF LANDSCAPE ARCHITECTURE

For the period before 1800, the history of landscape gardening (later called landscape architecture) is largely that of master planning and garden design for manor houses, palaces and royal properties, religious complexes, and centers of government. An example is the extensive work by André Le Nôtre at Vaux-le-Vicomte for King Louis XIV of France at the Palace of Versailles. The first person to write of *making* a landscape was Joseph Addison in 1712. The term landscape architecture was invented by Gilbert Laing Meason in 1828, and John Claudius Loudon (1783–1843) was instrumental in the adoption of the term landscape architecture by the modern profession. He took up the term from Meason and gave it publicity in his Encyclopedias and in his 1840 book on the *Landscape Gardening and Landscape Architecture of the Late Humphry Repton*.

Orangery at the Palace of Versailles, outside Paris

The practice of landscape architecture spread from the Old to the New World. The term "landscape architect" was used as a professional title by Frederick Law Olmsted in the United States in 1863 and Andrew Jackson Downing (1815–1852), another early American landscape designer, was editor of *The Horticulturist* magazine (1846–52). In 1841 his first book, *A Treatise on the Theory and Practice of Landscape Gardening, Adapted to North America*, was published to a great success; it was the first book of its kind published in the United States. During the latter 19th century, the term landscape architect begun to be used by professional landscapes designers, and was firmly established after Frederick Law Olmsted, Jr. and Beatrix Jones (later Farrand) with others founded the American Society of Landscape Architects (ASLA) in 1899. IFLA was founded at Cambridge, England, in 1948 with Sir Geoffrey Jellicoe as its first president, representing 15 countries from Europe and North America. Later, in 1978, IFLA's Headquarters were established in Versailles.

RELATION TO URBAN PLANNING

The combination of the traditional landscape gardening and the emerging city planning combined together gave landscape architecture its unique focus.Frederick Law Olmsted used the term 'landscape architecture' using the word as a profession for the first time when designing the Central Park.

Through the 19th century, urban planning became a focal point and central issue in cities. The combination of the tradition of landscape gardening and the emerging field of urban planning offered Landscape Architecture an opportunity to serve these needs. In the second half of the century, Frederick Law Olmsted completed a series of parks which continue to have a huge influence on the practices of Landscape Architecture today.

Among these were Central Park in New York City, Prospect Park in Brooklyn, New York and Boston's Emerald Necklace park system. Jens Jensen designed sophisticated and naturalistic urban and regional parks for Chicago, Illinois, and private estates for the Ford family including Fair Lane and Gaukler Point. One of the original ten founding members of the American Society of Landscape Architects (ASLA), and the only woman, was Beatrix Farrand. She was design consultant for over a dozen universities including: Princeton in Princeton, New Jersey; Yale in New Haven, Connecticut; and the Arnold Arboretum for Harvard in Boston, Massachusetts. Her numerous private estate projects include the landmark Dumbarton Oaks in the Georgetown neighborhood of Washington, D.C.. Since that time, other architects — most notably Ruth Havey and Alden Hopkins— changed certain elements of the Farrand design.

Since this period Urban Planning has developed into a separate independent profession that has incorporated important contributions from other fields such as Civil Engineering, Architecture and Public Administration. Urban Planners are qualified to perform tasks independent of landscape architects, and in general, the curriculum of landscape architecture programs do not prepare students to become urban planners.

Landscape architecture continues to develop as a design discipline, and to respond to the various movements in architecture and design throughout the 20th and 21st centuries. Thomas Church was a mid-century landscape architect significant in the profession. Roberto Burle Marx in Brazil combined the International style and native Brazilian plants and culture for a new aesthetic. Innovation continues today solving challenging problems with contemporary design solutions for master planning, landscapes, and gardens.

Ian McHarg was known for introducing environmental concerns in landscape architecture. He popularized a system of analyzing the layers of a site in order to compile a complete understanding of the qualitative attributes of a place. This system became the foundation of today's Geographic Information Systems (GIS). McHarg would give every qualitative aspect of the site a layer, such as the history, hydrology, topography, vegetation, etc. GIS software is ubiquitously used in the landscape architecture profession today to analyze materials in and on the Earth›s surface and is similarly used by Urban Planners, Geographers, Forestry and Natural Resources professionals, etc.

PROFESSION

In many countries, a professional institute, comprising members of the professional community, exists in order to protect the standing of the profession and promote its interests, and sometimes also regulate the practice of landscape architecture. The standard and strength of legal regulations governing landscape architecture practice varies from nation to nation, with some requiring licensure in order to practice; and some having little or no regulation. In North America, Europe, Australia and New Zealand, landscape architecture is a regulated profession.

Australia

The Australian Institute of Landscape Architects (AILA) provides non statutory professional recognition for landscape architects. Once recognized by AILA, landscape architects use the title 'Registered Landscape Architect'. Across the eight states and territories within Australia, there is a mix of requirements for landscape architects to be 'Registered', however it is not always a statutory requirement to be registered with AILA to practice use the term "Landscape Architect".

Any regulations or requirements are state based, not national. The AILA's system of professional recognition is a national system overseen by AILA's National Office in Canberra. Non (A.I.L.A) Landscape Architects are professionals who are also paid to undertake a specialised set of tasks and to complete them for a fee.

Some agencies require AILA professional recognition or registration as part of the pre-requisite for contracts. Landscape architects within Australia find that some contracts and competitions require the AILA recognition or 'registration' as the basis of demonstrating a professional status. To apply for AILA Registration, an applicant usually needs to satisfy a number of pre-requisites, including university qualification, two years of practice and a record of continuing professional practice. The application is in two stages: (1) A minimum 12 months of mentoring and assessment; and (2) oral assessment/ interview. Professional recognition includes a commitment to continue professional development. A.I.L.A Registered Landscape Architects are required to report annually on their continuing professional development.

Landscape Architecture within Australia covers a broad spectrum of design, advice and research. From specialist design services for commercial and government developments through to professional advice as an expert witness, the range of tasks delivered by Australian Landscape Architect's is diverse and interesting.

The harsh Australian environment also provide numerous challenges that must be overcome. Australia is the driest inhabited continent in the world and this characteristic determines particular requirements through design including specific species selection and careful consideration of natural resources such as rainfall and topography.

Canada

In Canada, landscape architecture, like law and medicine, is a self-regulating profession pursuant to provincial statute. For example, Ontario's profession is governed by the

Ontario Association of Landscape Architects pursuant to the *Ontario Association of Landscape Architects Act*. Landscape architects in Ontario, British Columbia, and Alberta must complete the specified components of L.A.R.E (Landscape Architecture Registration Examination) as a prerequisite to full professional standing.

Provincial regulatory bodies are members of a national organization, the Canadian Society of Landscape Architects / L'Association des Architectes Paysagistes du Canada (CSLA-AAPC), and individual membership in the CSLA-AAPC is obtained through joining one of the provincial or territorial components.

Italy

AIAPP (Italian Association of Landscape Architecture) is the Italian association of professional landscape architects formed in 1950 and is a member of IFLA and IFLA Europe (formerly known as EFLA). AIAPP is in the process of contesting this new law which has given the Architects' Association the new title of Architects, Landscape Architects, Planners and Conservationists whether or not they have had any training or experience in any of these fields other than Architecture. In Italy, there are several different professions involved in landscape architecture:

- Architects
- Landscape designers
- Doctor landscape agronomists and Doctor landscape foresters, often called Landscape agronomists.
- Agrarian Experts and Graduated Agrarian experts.

New Zealand

The New Zealand Institute of Landscape Architects (NZILA) is the professional body for Landscape Architects in NZ www.nzila.co.nz.

In April 2013, NZILA jointly with AILA, hosted the 50th International Federation of Landscape Architects (IFLA) World Congress in Auckland, New Zealand. The World Congress is an international conference where Landscape Architects from all around the globe meet to share ideas around a particular topic.

Within NZ, Members of NZILA when they achieve their professional standing, can use the title Registered Landscape Architect NZILA.

NZILA provides an education policy and an accreditation process to review education programme providers; currently there are three accredited undergraduate Landscape Architecture programmes in New Zealand. Lincoln University also has an accredited masters programme in landscape architecture.

Republic of Ireland

The professional body in Ireland for landscape architects is the Irish Landscape Institute (ILI) www.irishlandscapeinstitute.com. The ILI is an affiliate body to the European

Federation for Landscape Architecture (EFLA) and IFLA. The ILI was formed in 1993 to merge the disciplines of landscape architecture and landscape horticulture. It continues to promote the profession by its accreditation of the degree programme in Dublin, certification of Continuing Professional Development (CPD) for landscape architects, administration of professional practice examinations, advice on development of policy at national level and organisation of conferences, lectures and design awards. The ILI is a member institute of the Urban Forum, representing professional bodies involved in urban spatial disciplines of engineering, architecture, planning, quantity surveying and landscape architecture.

The profession has gained in status and numbers due to the construction boom of the past decade and raising of standards of Irish design. There is still no registration of title in Ireland and the profession is unregulated, but there is increasing awareness of the profession and of status of the ILI. Landscape architects in Ireland work in private practice, public sector bodies at local government level and in some bodies such transport and national heritage and in the academic sector. The demand for landscape architects is often associated with strategic infrastructure projects due to Ireland›s recent major infrastructural investments. Landscape architects are employed in design of: green infrastructure, public realm, institutional/medical/industrial campuses and settings, parks, play facilities, transport (road/rail/cycle/port) corridors, retail complexes, residential estates (including plans for remediation of now-abandoned housing ‹ghost› estates), village improvements, accessibility audits, graveyard restoration schemes, wind farms, wetland drainage systems and coastal zones. They are also significantly employed in preparation/review of statutory impact assessment reports on landscape, visual and ecological impacts of design proposals.

South Africa

In May 1962, Joane Pim, Ann Sutton, Peter Leutscher and Roelf Botha (considered the forefathers of the profession in South Africa) established the Institute for Landscape Architects, now known as the Institute for Landscape Architecture in South Africa (ILASA). ILASA is a voluntary organisation registered with the South African Council for the Landscape Architectural Profession. It consists of three regional bodies namely, Gauteng, KwaZula-Natal and the Western Cape. ILASA's mission is to advance the profession of landscape architecture and uphold high standards of professional service to its members, and to represent the profession of landscape architecture in any matter which may affect the interests of the members of the Institute. ILASA holds the country's membership with The International Federation of Landscape Architects (IFLA).

In South Africa, the profession is regulated by The South African Council for the Landscape Architectural Profession (SACLAP), established as a statutory council in terms of Section 2 of the South African Council for the Landscape Architectural Profession Act – Act 45 of 2000. The Council evolved out of the Board of Control for Landscape Architects (BOCLASA), which functioned under the Council of Architects in terms of The Architectural Act, Act 73 of 1970. SACLAP's mission is to establish, direct, sustain and ensure a high level of professional responsibilities and ethical conduct within the art and science of landscape architecture with honesty, dignity and integrity in the broad interest of public health, safety and welfare of the community. After completion of an accreditted under-graduate and/or post-graduate qualification in landscape architecture at either the University of Cape Town or the University of Pretoria, or landscape technology at the Cape Peninsula University of Technology, professional registration is attained via a mandatory mentored candidacy period (minimum of two years) and sitting of the professional registration exam. After successfully completing the exam, the individual is entitled to the status of Professional Landscape Architect or Professional Landscape Technologist.

United Kingdom

The UK's professional body is the Landscape Institute (LI). It is a chartered body which accredits landscape professionals and university courses. At present there are fifteen accredited programmes in the UK. Membership of the LI is available to students, academics and professionals, and there are over 3,000 professionally qualified members.

The Institute provides services to assist members including support and promotion of the work of landscape architects; information and guidance to the public and industry about the specific expertise offered by those in the profession; and training and educational advice to students and professionals looking to build upon their experience.

In 2008, the LI launched a major recruitment drive entitled "I want to be a Landscape Architect" to encourage the study of Landscape Architecture. The campaign aims to raise the profile of landscape architecture and highlight its valuable role in building sustainable communities and fighting climate change.

United States

In the United States, Landscape Architecture is regulated by individual state governments. For a landscape architect, obtaining licensure requires advanced education and work experience, plus passage of the national examination. Several states require passage of a state exam as well. In the United States licensing is overseen both at the state level, and nationally by the Council of Landscape Architectural Registration Boards (CLARB). Landscape architecture has been identified as an above-average growth profession by the US Bureau of Labor Statistics and was listed in *U.S. News & World Report*'s list of Best Jobs to Have in 2006, 2007, 2008, 2009 and 2010. The national trade association for United States landscape architects is the American Society of Landscape Architects.

EXAMPLES OF LANDSCAPE ARCHITECTURE

Potager in Dordogne, France

Monastery garden Bebenhausen Abbey in Baden-Württemberg, Germany

Entrance to walled Kitchen garden in Norfolk, England

Medieval garden (reconstruction) in Île-de-France, France

Woodland in Saihō-ji, Japan

Japanese garden in Ōtsu, Japan

Classical Chinese garden

Garden lighting in Pinjore, India

Topiary in Helsingborg, Sweden

Asian Sculpture garden in Texas, United States

Vigeland Sculpture garden in Oslo, Norway

Roof terrace garden (Ventimiglia, Italy)

Escorial Formal palace garden in Madrid, Spain

Mediterranean garden in Alpes-Maritimes, France

Herb garden in Warwickshire, England

Use of steps at Villa la Magia, in Quarrata, Italy.

REFERENCES

1. Hyams, Edward (1971). A History of Gardens and Gardening. New York, Washington: Praeger Publishers. p. 239.
2. Sir Geoffrey Jellicoe, Susan Jellicoe, The Landscape of Man: Shaping the

Environment from Prehistory to the Present Day ISBN 9780500274316

3. "Landscape Architecture - Your Environment. Designed". Asla.org. Retrieved 2013-04-06.
4. "Extensive Vegetative Roofs | Whole Building Design Guide". www.wbdg.org. Retrieved 2015-12-28.
5. London: Longman.
6. Find a Grave
7. "History of Horticulture". Ohio State University. Retrieved July 24, 2012.
8. IFLA Past, Present, Future - A publication about the history of IFLA. ISBN 3-9522080-0-0
9. UNCESCO Documents and Publications
10. The International Federation of Landscape Architects (IFLA)
11. Van Assche, K., Beunen, R., Duineveld, M., & de Jong, H. (2013). Co-evolutions of planning and design: Risks and benefits of design perspectives in planning systems. Planning Theory, 12(2), 177-198.
12. National Park Service (2000). Cultural Landscape Report: Dumbarton Oaks Park, Rock Creek Park. Washington, D.C.: U.S. Dept. of the Interior.
13. "Bulletin of Information for the AICP Comprehensive Planning Examination" (PDF). www.planning.org. Retrieved 2016-08-29. There are important distinctions between planners and allied professionals and between planning and related fields. Planners approach problems comprehensively, have a long-range perspective, and deal with unique place-based issues. Although people in related professions (e.g., law, architecture, landscape architecture, engineering, real estate development, etc.) and disciplines (humanities, psychology, etc.) often work with planners, they do not necessarily have the same knowledge base, skillset, and approach.
14. Corbett, John. "Ian McHarg: Overlay Maps and the Evaluation of Social and Environmental Costs of Land Use Change". Center for Spartially Integrated Social Science. Retrieved 4 June 2014.
15. Ozio, Ron (6 March 2001). "Obituary: Ian McHarg Dies". Penn News. University of Pennsylvania. Retrieved 4 June 2014.
16. "The Ontario Association of Landscape Architects". Oala.on.ca. Retrieved 2013-04-06.
17. Australian Institute of Landscape Architects (AILA)
18. Canadian Society of Landscape Architects Official Site - English
19. http://www.ilasa.co.za/
20. http://www.saclap.org.za/
21. "CC Position Statement". landscapeinstitute.org. 2011. Retrieved 30 May 2011.
22. "Career Advice and Guide for Job Searches - US News Business". Money.usnes.com. Retrieved 2013-04-06.

CHAPTER 5

Tools For Gardening

Chapter Contents

- **Introduction**
- **Hand Tools**
- **String Trimmer**
- **Diamond Tool**

Even among the basic garden tools, you are bound to find a wide variety of designs and sizes - enough to cause confusion when shopping. There is a reason for so many options - using the right tool for the right job makes your work easier and more efficient.

INTRODUCTION

A garden tool is any one of many tools made for gardens and gardening and overlaps with the range of tools made for agriculture and horticulture. Garden tools can also be hand tools and power tools.

A collection of various garden tools.

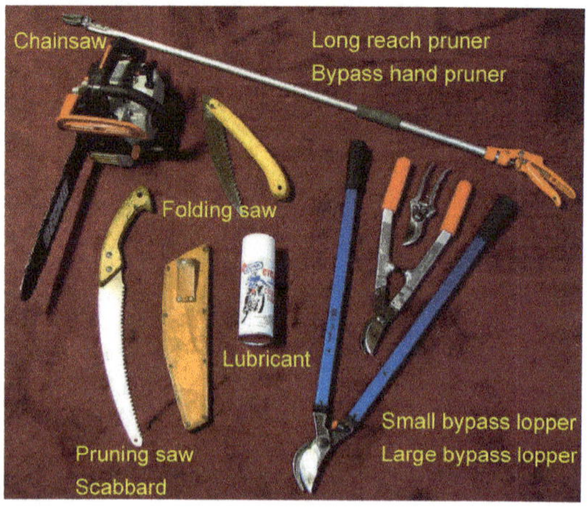

Some pruning tools that can be used to maintain a garden.

HAND TOOLS

The hand tools still used by gardeners originated with the earliest agricultural implements used by humans. Examples include:

- Axe
- Sickle
- Pitchfork
- Spade
- Scythe
- Shovel
- Trowel
- Hoe
- Fork
- Rake
- machete as well.

The earliest tools were made of wood, flint, metal, tin, and bone. However, the development of metal working, first in copper and later in iron and steel, enabled the manufacture of more durable tools. Industrial metalworking enabled the manufacture of efficient cutting tools, including pruning shears (secateurs – for example anvil pruning shears), grass shears, loppers and larger, more efficient powered tools such as the trencher.

Ergonomics

Modern tool design includes ergonomics considerations. Tools are being designed that induce ever less stress on the human body when used. The most efficient tools keep the body in a neutral position while being used. This helps to lessen the stress on joints and muscles. An advantage of this approach is that it requires gardeners to exert less energy whilst using the tools. Some modern patents such as US 7,832,125 show an example without hand stop provisions.

Power Tools

The first power tool to become popular with gardeners was the lawn mower. This has been followed by a very wide range of power tools, including :

- Cultivators
- String trimmer
- Irrigation sprinklers
- Hedge trimmers
- Lawn aerators
- Leaf sweepers
- Trenchers
- Leaf blowers
- Chainsaws
- Mini-tractors

Environmental Impact

The exhaust fumes from gas-powered equipment is a significant source of air pollution. US emission standards specifically limit emissions from these small engines.

STRING TRIMMER

A string trimmer, also called a "weed-whip", "whipper-snipper", "weed-whacker", a "weed eater", a "line trimmer" (in Australia) or (in the UK and Ireland) a "strimmer", is a tool which uses a flexible monofilament line instead of a blade for cutting grass and other plants near objects, or on steep or irregular terrain. It consists of a cutting head at the end of a long shaft with a handle or handles and sometimes a shoulder strap.

A man using a gasoline-powered string trimmer

Design

A sample of string trimmer heads. From left: A string or line head, a chain trimmer head (now banned in the EU after a fatality), and a metal brush cutting head.

A string trimmer works on the principle that a line that is turned fast enough is held out from its housing (the rotating reel) very stiffly by centrifugal force. The faster it turns the stiffer the line. Even round-section nylon line is able to cut grass and slight, woody plants quite well. Some monofilament lines designed for more powerful cutters have an extruded shape, like a star, that helps the line slash the material being cut; the line is thus able to cut quite large woody plants (small shrubs) or at least ring-bark them very effectively. These lines make disks less necessary for tough jobs.

The line is hand-wound onto a reel before the job is started, leaving both ends extending from the reel housing. The motor turns the reel and the line extends horizontally while the operator swings the trimmer about where the plants are to be trimmed. The operator controls the height at which cutting takes place and can trim down to ground level quite easily.

As the line is worn, or breaks off, the operator knocks the reel on the ground so that a release mechanism allows some of the line in the reel to extend and replace the spent portion. The newer models have an 'auto-feed' operation where a small cutter on the line-guard ensures that the line length exposed for cutting does not exceed the length that can be swung efficiently by the motor. Newly extended line operates more efficiently because of its heavier weight and surface effects. The speed of the spinning hub is usually controlled by a trigger on the handle.

For vertical cutting the whole machine can be tilted or some trimmers allow the head to be adjusted at different angles. Vertical cutting is not recommended near sidewalks or other concrete and pavement edges, because it leaves open grooves that allow water to collect and cause damage.

String trimmers powered by an internal combustion engine have the engine on the opposite end of the shaft from the cutting head, while electric string trimmers typically have an electric motor in the cutting head, but some other arrangements exist too. One of such is an arrangement where the trimmer is connected to heavy machinery and is powered using a hydraulic motor.

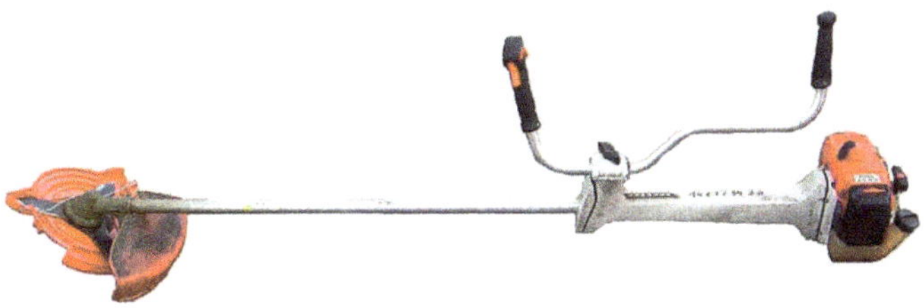

A brush cutter

The head contains a safety shield on the user side and a rotating hub which may also be called a head or spool. Disadvantages of a gasoline-powered string trimmer include its greater weight and the significant vibration that carries throughout the device, both of which interfere with its maneuverability and contribute to muscle fatigue, as well as the requirement that motor oil be added to its fuel (if it is equipped with a two-stroke engine). Advantages of gasoline-powered trimmers include mobility (because they are not attached to a power outlet) and the higher maximum power. Large trimmers, used for cutting roadside grass in large areas, are often heavy enough to require two hands to operate, and some are even fitted with a harness enabling the user's torso to bear some of their weight. These very large trimmers are often referred to as brush cutters. Brush-cutter types are usually made so that a metal blade can be attached instead of the "string" (or monofilament). A metal blade enables cutting heavier woody brush.

Trimmers that have nylon or metal blades usually require straight driveshafts to handle the higher torque required to turn the heavier disk, and because of the shock loads that are passed back from the blade to the drive shaft and its gearbox(es). Smaller line trimmers have curved driveshafts to make holding the cutting-head at ground level much easier and with less strain on the operator.

Accessories

Many string trimmers allow the hub, the head or the lower part of the shaft to be replaced with accessories. Common accessories include:

- replacing the monofilament line with metal or plastic blades.
- replacing the lower shaft with a small chain saw to create a powered pole saw.
- replacing the lower shaft with a hedge trimmer.
- replacing the lower shaft with a cultivator.

Quick-release shafts are offered on many newer models which do not require any tools to switch in accessories.

Power and Emissions

Gasoline-engine powered trimmers usually have a minimum of 21 cc displacement motors. At this size they can easily turn 2-millimetre (0.079 in) line and some have nylon blades as accessories to the line-reel. A 32 cc engine can swing a 2.75 mm (0.108 in) line and often has metal-blade accessories. Most trimmers use two stroke engines and require gasoline mixed with oil. Due to pollution laws four stroke engines are becoming more popular. For instance, Honda, MTD and Craftsman manufacture a four stroke engine trimmer. Other companies, such as John Deere, now carry low-emission two-stroke engine trimmers. Stihl manufactures a hybrid four stroke engine trimmer using a technology called 4MIX. 4MIX trimmers have no oil reservoir. This engine is lubricated using pre-mixed gasoline, like a two-stroke engine.

Electric edge trimmers have the advantage of being very light, easy to maneuver and easy-to-operate devices. However, the length of power cord that can be deployed across the ground limits them, and they are usually less powerful and robust than the gasoline-engine ones. Electric machines normally are limited to 2.5 mm (0.098 in) maximum diameter nylon because of their lower power output (400 to about 1200 watts). Recently there are electric string trimmers that offer the same performance as gasoline-powered trimmers. Battery-powered units are to be recharged after each use. The recharge time is typically several hours long; some models offer a quick-charge option that cuts charging time to between 30 minutes and 2 hours, or a removable battery pack so the user can have more than one battery ready to swap out when the first one runs down.

The typical two-cycle engine used on string trimmers pollutes heavily due to incomplete combustion in the cylinder. This results in unburned fuel escaping through the exhaust system and into the environment. US emission standards specifically limit emissions from small engines. Electric models produce no emissions at the point of use. Battery-powered units typically use small or large sealed lead acid, nickel metal hydride, or lithium ion batteries.

Safety Concerns

String trimmers can be dangerous tools due to the fact that they can cause debris, including rocks and stones, to go flying in several directions. For this reason, it is typical for the person who is using a string trimmer to wear either safety glasses or a visor to protect the face and particularly the eyes as a stone in the eye could cause a person to be blinded for life. This, however, does not offer any protection for passersby, or nearby people who may also be vulnerable to being hit by debris from a string trimmer. In addition to the safety risk posed to individual people, the debris from string trimmer also has the potential to cause damage to cars and buildings, with a particularly high risk of breaking glass on windows.

Chain-link flail rotors, and any other trimmer head with linked metal parts are prohibited from sale in Britain and EU member states following their identifiation as a serious safety risk. This follows a fatal accident in 2010 where a chain section became detached from such a rotor and struck a nearby worker at sufficient speed to cause a lethal injury.

Other use of String Trimmer Parts

- The engine from a string trimmer may be used to power DIY motorized bicycles.
- String trimmer engines may be used to power radio-controlled models.
- Round shaped 0.65mm trimmer lines can be use as replacement nibs for Wacom drawing tablet pens.

DIAMOND TOOL

A diamond tool is a cutting tool with diamond grains fixed on the functional parts of the tool via a bonding material or another method. As diamond is a superhard material, diamond tools have many advantages as compared with tools made with common abrasives such as corundum and silicon carbide.

A close-up of a segment of a diamond saw blade

Advantages

Diamond is the hardest material on the earth; much harder than corundum and silicon carbide. Diamond also has high strength, good wear resistance, and a low friction coefficient. So when used as an abrasive, it has many obvious advantages over many other common abrasives.

Advantages of Diamond Grinding Tools

Diamond can be used to make grinding tools, which have following advantages:

- **High grinding efficiency, Low grinding force**: Less heat will be generated by the hole in the grinding process. This can decrease or avoid burns and cracks on the surface of the workpiece, and decrease the equipment's wear and energy consumption.
- **High wear resistance**: Diamond grinding tools' change in dimension is small. This can lead to good grinding quality and high grinding precision.
- **Long lifespan, Long dressing period**: This can greatly increase the work efficiency, and improve the workers' labor environment and decrease the product's labor intensity.
- **Low comprehensive cost**: The processing cost of each workpiece is lower.

Categories

There are thousands of kinds of diamond tools. They can be categorized by their manufacturing methods and their uses.

Categories by Manufacturing Method

According to their manufacturing methods or bond types, diamond tools can be categorized to the following way:

- **Metal-bonded diamond tools**: The tools' bonding material is metal mixture powder. The functional parts of the tool are usually diamond segments. These tools include metal-bonded diamond saw blades, diamond grinding cup wheels, diamond core drill bits, etc. For metal-bonded diamond tools, the bond is one of the prime factors when selecting which tool to use for cutting or grinding a specific material, depending on how hard or abrasive the material is. The bond used dictates the rate at which the metallic powders wear down and expose new diamond crystals at the surface, thereby maintaining an abrasive cutting surface. Different bond strengths are achieved by the alloy mix of metallic powders chosen and how much heat and pressure are applied to the sintered segment.
- **Resin-bonded diamond tools**: The tools' bonding material is mainly resin powder. An example of this tool is the resin-bonded diamond polishing pads used in the construction industry.
- **Plated diamond tools**: This tool is made by fixing the diamonds onto the tool's base via electroplating method or via CVD (Chemical Vapor Deposition) method. This tool can usually be made to good processing precision.
- **Ceramic-bonded diamond tools**: The tools' bonding material is usually glass and ceramic powder. This tool usually has the features of good chemical stability, small elastic deformation, but high brittleness, etc.
- **Polycrystalline Diamond (PCD)**: They are normally made by sintering many micro-size single diamond crystals at high temperature and high pressure. PCD has good fracture toughness and good thermal stability, and is used in making geological drill bits.
- **Polycrystalline Diamond Composite or Compacts (PDC)**: They are made by combining some layers of polycrystalline diamonds (PCD) with a layer of cemented carbide liner at high temperature and high pressure. PDC has the advantages of diamond's high wear resistance with carbide's good toughness.
- **High-temperature brazed diamond tools**: This tool is made by brazing a single layer of diamonds onto the tool via a solder at a temperature of over 900 °C. This tool is a newly developed product. Its manufacture uses technologies include vacuum brazing and atmosphere-protected brazing. This tool has several advantages: the solder can hold the diamonds very firmly, the single layer of diamonds' exposed height can be 70%–80% of their sizes, and the diamonds can be regularly arranged on the tool.

Categories by Use

If categorized by use, there are diamond grinding tools, diamond cutting tools (e.g., diamond coated twist drill bits), diamond drilling tools, diamond sawing tools (e.g., diamond saw blades), diamond drawing dies, etc.

Applications

Applicable Materials

Diamond tools are suitable to process the following materials:

- Carbide alloy
- Hard or abrasive non-metallic materials, for example, stone, concrete, asphalt, glass, ceramics, gem stone and semiconductor materials.
- Non-ferrous metals such as aluminum, copper and their alloys, and some soft but tough materials such as rubber and resin.

As diamonds can react with Fe, Co, Ni, Cr, V under the high temperatures generated in the grinding processes, normally diamond tools are not suitable to process steels, including common steels and various tough alloy steels, while the other superhard tool, cubic boron nitride (CBN) tool, is suitable to process steels. The tools made with common abrasives (e.g. corundum and silicon carbide) can also do the task.

Applied Domains

Diamond tools are used in the following domains:

- **Geological or project exploration**: Diamond geological drill bits, diamond oil drill bits and diamond thin-wall drill bits are often used. The main application of PCD drilling bits is in the oil and natural gas industries and the mining industry.
- **Stone processing**: Diamond circular saw blades, diamond gang saws, diamond band saws are used to saw marble, granite and other stone blocks. Diamond wire saws are used in stone quarries to exploit raw stone blocks. Shaped diamond tools are used to process stone to a particular shape. Resin-bonded diamond polishing pads are used to polish stone.
- **Construction**: Medium or small sized diamond saw blades, diamond core drill bits and some diamond grinding or polishing tools are often used in repairing roads, remodeling buildings, and processing building materials.
- **Woodworking**: Composite laminate flooring is widely used. It is wearable as stone. PCD circular saw blades, profiling cutter, twist drill bits and other diamond tools are used in its processing.
- **Auto spare parts processing**: PCD and PCBN cutting tools are used to meet the high efficiency and low deviation processing requirements in this domain.
- **IT and home appliance products processing**: High-precision super-thin diamond cutting wheels are used to cut silicon slices. Resin-bonded diamond grinding

wheels are used to process ceramics in optical fiber industry.

- **Engineering ceramics processing**: Engineering ceramics are widely used in many industries. They have the properties of high toughness, high hardness, high-temperature resistance. High-toughness and durable diamond grinding wheels are developed to process them.
- **Carbide tools and other mechanical tools processing**: Diamond tools are used to gain high processing precision and efficiency.

Besides what are listed above, there are also other domains where diamond tools are applied, for example, in medicine, Venezuelan scientist Humberto Fernandez Moran invented the diamond knife for use in delicate surgeries in 1955.

Apart from its use as an abrasive due to its high hardness, diamond is also used to make other products for its many other good properties such as high heat-conductivity, low friction coefficient, high chemical stability, high resistivity and high optical performances. These applications include coatings on bearings and CDs, acting as lens and thermistors, making high-voltage switches and sensors, etc.

Some Examples of Diamond Tools

Diamond Dressing Tools

Diamond dressers consist of single-point or multipoint tools brazed to a steel shank, and used for the trueing and dressing of grinding wheels. The tools come in several types, including: grit impregnated, blade type, crown type, and disc type. The advantages of multipoint over single-point tools are:

- The whole diamond can be used; in a single-point tool, when the point is blunt the diamond must be reset, and after few resettings the diamond is replaced.
- Multipoint tools have higher accuracy, especially in form grinding, where blade types are used. Blades consist of elongated diamonds. The thickness is controlled and blades are available in thicknesses from 0.75 to 1.40 millimeters (0.030 to 0.055 in).
- Grit-type tools are of a tough grade, and can be used for bench grinders.
- Since small points are used, the diamonds have a cutting edge with natural points, unlike single-point tools, which have brutted points.
- The cost of multipoint tools is lower, since smaller, less expensive diamonds are used.

PCD Cutting Tools

Polycrystalline diamond (PCD) is formed in a large High Temperature-High Pressure (HT-HP) press, as either a diamond wafer on a backing of carbide, or forming a «vein» of diamond within a carbide wafer or rod.

Most wafers are polished to a mirror finish, then cut with an electrical discharge machining (EDM) tool into smaller, workable segments that are then brazed onto the sawblade, reamer, drill, or other tool. Often they are EDM machined and/or ground an additional time to expose the vein of diamond along the cutting edge. These tools are mostly used for the machining of nonmetallic and nonferrous materials.

The grinding operation is combined with EDM for several reasons. For example, according to Modern Machine Shop, the combination allows a higher material removal rate and is therefore more cost effective. Also, the EDM process slightly affects the surface finish. Grinding is used on the affected area to provide a finer final surface. The Beijing Institute of Electro-Machining attributes a finer shaping and surface geometry to the combination of the two processes into one.

The process itself is accomplished by combining the two elements from each individual process into one grinding wheel. The diamond graphite wheel accomplishes the task of grinding, while the graphite ring around the existing wheel serves as the EDM portion. However, sin

ce diamond is not a conductive material, the bonding in the PCD work piece must be ample enough to provide the conductivity necessary for the EDG process to work.

Polycrystalline diamond tools are used extensively in automotive and aerospace industries. They are ideal for speed machining (9000 surface feet per minute or higher) in tough and abrasive aluminum alloys, and high-abrasion processes such as carbon-fiber drilling and ceramics. The diamond cutting edges make them last for extended periods before replacement is needed. High volume processes, tight tolerances, and highly abrasive processes are ideal for diamond tooling.

Polycrystalline Diamond Compacts

In the late 1970s, General Electric pioneered the technology of polycrystalline diamond compacts (PDCs) as a replacement for natural diamonds in drill bits. PDCs have been used to cut through crystalline rock surfaces for extended periods of time in lab environments, and these capabilities have now been implemented in harsh environments throughout the world.

As of August 2000, the U.S. Department of Energy claimed that nearly one-third of the total footage drilled worldwide is being drilled with PDC bits, with a claimed savings of nearly $100,000 per PDC bit as compared to roller-core bits.

Diamond Paste and Slurry

Diamond pastes are used for polishing materials that require a mirror finish. They are often used in metallurgical specimens, carbide dies, carbide seals, spectacle glass industry, and for polishing diamonds. Diamond paste is mainly used in industrial requirements for polishing and sharpening metal blades and other metal surfaces. The paste is not just to polish the metal blade but sharpen the cutting edge as well.

Diamond Electroplated Tools

Diamond powder deposited through electroplating is used to make files (including nail files) and in small grinding applications.

Single Point Diamond Turning Tools

Single point diamond turning (SPDT) utilizes a solid, flawless diamond as the cutting edge. The single crystalline diamond can be natural or synthetic, and is sharpened to the desired dimensions by mechanical grinding and polishing. The cutting edge of most diamond tools is sharp to tens of nanometers, making it very effective for cutting non-ferrous materials with high resolution. SPDT is a very accurate machining process, used to create finished aspherical and irregular optics without the need for further polishing after completion. The most accurate machine tool in the world, the LODTM, formerly at Lawrence Livermore National Laboratory, had a profile accuracy estimated at 28 nm, while most machines seek a roughness within that deviation.

SPDT is used for optics, for flat surfaces where both surface finish and unusually high dimensional accuracy are required, and when lapping would be uneconomical or impractical.

REFERENCES

1. "Weed Eater inventor dies at 85", June 29, 2011, Associated Press, Lake Charles American Press
2. heavy duty lawn trimmer
3. "Use of chain flail/non standard cutting attachments on brush cutters". www.hse.gov.uk. Retrieved 2016-07-25.
4. "Anthony Robinson death: Firm fined over strimmer accident". BBC News. Retrieved 2016-07-25.
5. Pliny the Elder;. "Natural History". Translated by D.E. Eichholz (vol. 10).
6. Journal of Petroleum Technology. "Legends of Drilling" (PDF). pp. 50–55.
7. National Renewable Energy Laboratory (August 2000). Diamond-Cutter Drill Bits (PDF) (PDF). United States Department of Energy. p. 2. (DOE/GO 100098-482)
8. The World's Most Accurate Lathe, Lawrence Livermore National Laboratory, April 2001

CHAPTER 6

Practices of Olericulture

Chapter Contents

Olericulture is the applied life science of vegetable cultivation and production. Olericulture is usually associated with the cultivation of annual, herbaceous garden plants. Olericulture deals with herbaceous plants for the kitchen, including, for example, carrots (edible root), asparagus (edible stem), lettuce (edible leaf), cauliflower (edible flower), tomatoes (edible fruit), and peas (edible seed).

INTRODUCTION

Olericulture is the science of vegetable growing, dealing with the culture of non-woody (herbaceous) plants for food.

Olericulture is the production of plants for use of the edible parts. Vegetable crops can be classified into 9 major categories:

- Potherbs and greens – spinach and collards
- Salad crops – lettuce, celery
- Cole crops – cabbage and cauliflower
- Root crops (tubers) – potatoes, beets, carrots, radishes

- Bulb crops – onions, leeks
- Legumes – beans, peas
- Cucurbits – melons, squash, cucumber
- Solanaceous crops – tomatoes, peppers
- Sweet corn

Olericulture deals with the production, storage, processing and marketing of vegetables. It encompasses crop establishment, including cultivar selection, seedbed preparation and establishment of vegetable crops by seed and transplants.

It also includes maintenance and care of vegetable crops as well commercial and non-traditional vegetable crop production including organic gardening and organic farming; sustainable agriculture and horticulture; hydroponics; and biotechnology.

LEAF VEGETABLE

Leaf vegetables, also called potherbs, greens, vegetable greens, leafy greens, or salad greens, are plant leaves eaten as a vegetable, sometimes accompanied by tender petioles and shoots. Although they come from a very wide variety of plants, most share a great deal with other leaf vegetables in nutrition and cooking methods.

Nearly one thousand species of plants with edible leaves are known. Leaf vegetables most often come from short-lived herbaceous plants such as lettuce and spinach. Woody plants whose leaves can be eaten as leaf vegetables include *Adansonia*, *Aralia*, *Moringa*, *Morus*, and *Toona* species.

The leaves of many fodder crops are also edible for humans, but usually only eaten under famine conditions. Examples include alfalfa, clover, and most grasses, including wheat and barley. These plants are often much more prolific than more traditional leaf vegetables, but exploitation of their rich nutrition is difficult, primarily because of their high fiber content. This obstacle can be overcome by further processing such as drying and grinding into powder or pulping and pressing for juice.

Leaf vegetables contain many typical plant nutrients, but since they are photosynthetic tissues, their vitamin K levels in relation to those of other fruits and vegetables, as well as other types of foods, is particularly notable. The reason is that phylloquinone, the most common form of the vitamin, is directly involved in photosynthesis. This causes leaf vegetables to be the primary food class that interacts significantly with the anticoagulant pharmaceutical warfarin.

Spinach leaves in a colander.

Fresh Gotukola (Centella asiatica)

Fresh Swiss chard

Fresh water spinach

Sabzi Khordan, an Iranian salad-like dish.

Nutrition

Leaf vegetables are typically low in calories and fat, and high in protein per calorie, dietary fiber, vitamin C, pro-vitamin A carotenoids, folate, manganese and vitamin K.

The vitamin K content of leaf vegetables is particularly high, since these are photosynthetic tissues and phylloquinone is involved in photosynthesis. Accordingly, users of vitamin K antagonist medications, such as warfarin, must take special care to limit consumption of leaf vegetables.

Preparation

If leaves are cooked for food, they may be referred to as boiled greens. Leaf vegetables may be stir-fried, stewed, steamed, or consumed raw. Leaf vegetables stewed with pork are a traditional dish in soul food, and southern U.S. cuisine. They are also commonly eaten in a variety of South Asian dishes such as saag. Leafy greens can be used to wrap up other ingredients into an edible package similar to how a tortilla is sometimes used. Many green leafy vegetables, such as lettuce or spinach, can also be eaten raw, for example in sandwiches or salads. A green smoothie enables large quantities of raw leafy greens to be consumed by blending the leaves with fruit and water.

Africa

In certain countries of Africa, various species of nutritious amaranth are very widely eaten boiled.

Celosia argentea var. argentea or «Lagos spinach» is one of the main boiled greens in West African cuisine.

Called "liponda" in Chiyao, these are just greens cooked up and served usually with ugali

Greens to be cooked and accompany ugali in east Africa

This Yawo woman goes through a daily task of preparing greens for a meal

Greece

In Greek cuisine, *khorta* (χόρτα, lit. 'greens›) are a common side dish, eaten hot or cold and usually seasoned with olive oil and lemon.

At least 80 different kinds of greens are used, depending on the area and season, including: black mustard, dandelion, wild sorrel, chicory, fennel, chard, kale, mallow, black nightshade, lamb's quarters, wild leeks, hoary mustard, charlock, smooth sow thistle and even the fresh leaves of the caper plant.

Italy

Ligurian pansoti filled with *preboggion* boiled greens and served with nut sauce.

Preboggion, a mixture of different wild boiled greens, is used in Ligurian cuisine to stuff ravioli and pansoti. One of the main ingredients of *preboggion* are borage *(Borago officinalis)* leaves. *Preboggion* is also sometimes added to minestrone soup and frittata.

United States

In the cuisine of the southern United States and traditional African-American cuisine, turnip, collard, kale, garden cress, dandelion, mustard, and pokeweed greens are commonly cooked, and often served with pieces of ham or bacon. The boiling water, called potlikker, is used as broth. As water in which pokeweed (poke sallet, poke salad) is boiled contains toxins removed by the boiling, this water should be discarded.

SWEET CORN

Sweet corn (Zea mays convar. saccharata var. rugosa; also called sugar corn and pole corn) is a variety of maize with a high sugar content. Sweet corn is the result of a naturally occurring recessive mutation in the genes which control conversion of sugar to starch inside the endosperm of the corn kernel. Unlike field corn varieties, which are harvested when the kernels are dry and mature (dent stage), sweet corn is picked when immature (milk stage) and prepared and eaten as a vegetable, rather than a grain. Since the process of maturation involves converting sugar to starch, sweet corn stores poorly and must be eaten fresh, canned, or frozen, before the kernels become tough and starchy.

Husked sweet corn

Loose kernels of sweet corn

Anatomy

The fruit of the sweet corn plant is the corn *kernel*, a type of fruit called a caryopsis. The *ear* is a collection of kernels on the *cob*. Because corn is a monocot, there is always an even number of rows of kernels. The ear is covered by tightly wrapped leaves called the *husk*. *Silk* is the name for the pistillate flowers, which emerge from the husk. The husk and silk are removed by hand, before boiling but not necessarily before roasting, in a process called *husking* or *shucking*.

Consumption

In most of Latin America, sweet corn is traditionally eaten with beans; each plant is deficient in an essential amino acid that happens to be abundant in the other, so together sweet corn and beans form a protein-complete meal. In Brazil, sweet corn cut off from the cobs is generally eaten with peas (where this combination, given the practicality of steamed canned grains in an urban diet, is a frequent addition to diverse meals such as salads, stews, seasoned white rice, risottos, soups, pasta, and, most famously, whole sausage hot dogs).

Similarly, sweet corn in Indonesia is traditionally ground or soaked with milk, which makes available the B vitamin niacin in the corn, the absence of which would otherwise lead to pellagra; in Brazil, a combination of ground sweet corn and milk is also the basis of various well-known dishes, such as *pamonha* and the pudding-like dessert *curau*, while sweet corn eaten directly off the cobs tends to be served with butter.

The kernels are boiled or steamed. In Europe, China, Korea, Japan and India, they are often used as a pizza topping, or in salads. Corn on the cob is a sweet corn cob that has been boiled, steamed, or grilled whole; the kernels are then eaten directly off the cob or cut off. Creamed corn is sweet corn served in a milk or cream sauce. Sweet corn can also be eaten as baby corn.

In the United States, sweet corn is eaten as a steamed vegetable, or on the cob, usually served with butter and salt. It can be found in Tex-Mex cooking in chili, tacos, and salads. When corn is mixed with Lima beans it is called Succotash.Sweet corn is one of the most popular vegetables in the United States. fresh, canned and frozen sweetcorn rank among the top ten vegetables in value and per capital consumption.Frozen cut corn is exceed only by frozen potato products while frozen corn on the cob is 4th following peas.

If left to dry on the plant, kernels may be taken off the cob and cooked in oil where, unlike popcorn, they expand to about double the original kernel size and are often called corn nuts. A soup may also be made from the plant, called sweet corn soup.

Young sweet corn

The same rows of corn 41 days later at maturity.

Health Benefits

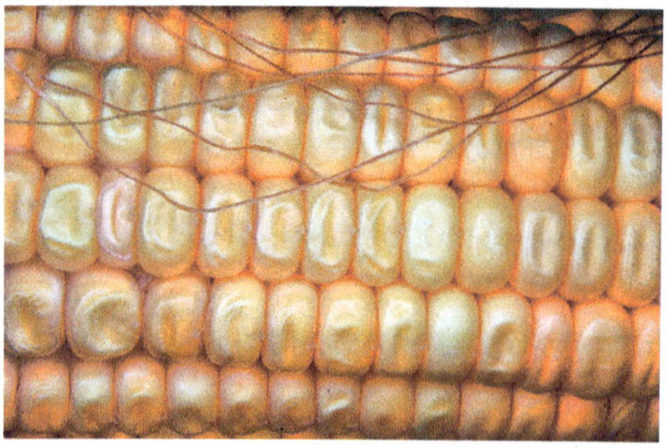

Overripe sweet corn

Cooking sweet corn increases levels of ferulic acid, which has anti-cancer properties.

Cultivars

Open pollinated (non-hybrid) corn has largely been replaced in the commercial market by sweeter, earlier hybrids, which also have the advantage of maintaining their sweet flavor longer. *Su* cultivars are best when cooked within 30 minutes of harvest. Despite their short storage life, many open pollinated cultivars such as 'Golden Bantam' remain popular for home gardeners and specialty markets, or are marketed as heirloom seeds. Although less sweet, they are often described as more tender and flavorful than hybrids.

Genetics

Early cultivars, including those used by Native Americans, were the result of the mutant *su* ("sugary") allele.They contain about 5–10% sugar by weight.

Supersweet corn are cultivars of sweet corn which produce higher than normal levels of sugar developed by University of Illinois at Urbana–Champaign professor John Laughnan. He was investigating two specific genes in sweet corn, one of which, the *sh2* gene, caused the corn to shrivel when dry. After further investigation Laughnan discovered that the endosperm of *sh2* sweet corn kernels store less starch and from 4 to 10 times more sugar than normal *su* sweet corn. He published his findings in 1953, disclosing the advantages of growing supersweet sweet corn, but many corn breeders lacked enthusiasm for the new supersweet corn. Illinois Foundation Seeds Inc. was the first seed company to release a supersweet corn and it was called ‹Illini Xtra Sweet›, but widespread use of supersweet hybrids did not occur until the early 1980s. The popularity of supersweet corn rose due to its long shelf life and large sugar content when compared to conventional sweet corn. This has allowed the long-distance shipping of sweet corn and has enabled manufacturers to can sweet corn without adding extra sugar or salt.

Cut White Sweet Corn

The third gene mutation to be discovered is the *se* or "sugary enhanced" allele, responsible for so-called "Everlasting Heritage" cultivars, such as 'Kandy Korn›. Cultivars with the *se* alleles have a longer storage life and contain 12–20% sugar compared to *su* cultivars.

All of the alleles responsible for sweet corn are recessive, so it must be isolated from other corn, such as field corn and popcorn, that release pollen at the same time; the endosperm develops from genes from both parents, and heterozygous kernels will be tough and starchy. The *se* and *su* alleles do not need to be isolated from each other. However supersweet cultivars containing the *sh2* allele must be grown in isolation from other cultivars to avoid cross-pollination and resulting starchiness, either in space (various sources quote minimum quarantine distances from 100 to 400 feet or 30 to 120 m) or in time (i.e., the supersweet corn does not pollinate at the same time as other corn in nearby fields).

Modern breeding methods have also introduced cultivars incorporating multiple gene types:

- *sy* (for *synergistic*) adds the *sh2* gene to some kernels (usually 25%) on the same cob as a *se* base (either homozygous or heterozygous)
- *augmented sh2* adds the *se* and *su* gene to a *sh2* parent

Often seed producers of the *sy* and *augmented sh2* types will use brand names or trademarks to distinguish these cultivars instead of mentioning the genetics behind them. Generally these brands or trademarks will offer a choice of white, bi-color and yellow cultivars which otherwise have very similar characteristics.

Bt Corn

Bt corn is genetically modified to resist certain insects. Commercial growers have access to sweet corn seed bred with this artificial trait. Bt corn and other transgenic varieties are not available to the home grower due to protocols that must be followed in their production.

REFERENCES

1. "Nutrition facts for raw spinach per 100 g; USDA Nutrient Data SR-21". Conde Nast. 2014. Retrieved 15 Sep 2014.
2. Kessler, F.; Glauser, G. (2014). "Prenylquinone Profiling in Whole Leaves and Chloroplast Subfractions". Plant Isoprenoids. Methods in Molecular Biology. 1153. p. 213. doi:10.1007/978-1-4939-0606-2_15. ISBN 978-1-4939-0605-5.
3. Chang, C. -H.; Wang, Y. -W.; Yeh Liu, P. -Y.; Kao Yang, Y. -H. (2014). "A practical approach to minimize the interaction of dietary vitamin K with warfarin". Journal of Clinical Pharmacy and Therapeutics. 39 (1): 56–60. doi:10.1111/jcpt.12104. PMID 24383939.
4. National Research Council (U.S.), Board on Science and Technology for International Development, Lost Crops of Africa: Vegetables, pp. 6, 35f. Books
5. "ECHO". ECHO.
6. Lynn Livanos Athan. "Horta Vrasta - Boiled Leafy Greens". About.com Food.
7. "dall'olio al preboggion". cogornese.it.
8. "Liguria-style omelette with borage and sausage". Italian recipes - Italian food culture - Academia Barilla.
9. Erwin, A. T. (July 1951). "Sweet Corn—Mutant or historic species?". Economic Botany. Springer New York. 5 (3): 302. doi:10.1007/bf02985153.
10. Schultheis, Jonathan R. "Sweet Corn Production." North Carolina Cooperative Extension Service, North Carolina State University. Revised 12/94.
11. "Stewart's Wilt of Corn". Retrieved 2014-07-07.
12. Levey Larson, Debra (August 2003). "Supersweet sweet corn: 50 years in the making". Inside Illinois. University of Illinois at Urbana-Champaign. 23 (3). Retrieved 2009-09-03.
13. Mann, Charles Kellogg. 1491: New Revelations of the Americas Before Columbus. Vintage. p. 221. ISBN 1-4000-3205-9.
14. "Plant breeding reviews volume 14".
15. "Cooking sweet corn boosts its ability to fight cancer and heart disease by freeing healthful compounds, Cornell scientists find". Cornell News. Retrieved 2009-09-07.
16. "Insect Resistance Management Fact Sheet For Bt Corn". National Corn Growers Association. Retrieved 2015-06-03.

CHAPTER 7

Viticulture

Chapter Contents

Viticulture is the study of the science of growing grapes, with the ones used to produce wine being the primary focus. It includes the study of vine selection, irrigation styles, planting techniques, pest and disease management, optimal harvesting and more. Anything that can possibly happen to the grape before the wine making process begins is included in viticulture.

INTRODUCTION

Viticulture is the science, production, and study of grapes. It deals with the series of events that occur in the vineyard. It is a branch of the science of horticulture.

While the native territory of Vitis vinifera, the common grape vine, ranges from Western Europe to the Persian shores of the Caspian Sea, the vine has demonstrated high levels of adaptability to new environments. For this reason, viticulture can be found on every continent except Antarctica.

Duties of the viticulturist include monitoring and controlling pests and diseases, fertilizing, irrigation, canopy management, monitoring fruit development and characteristics, deciding when to harvest, and vine pruning during the winter months. Viticulturists are often intimately involved with winemakers, because vineyard management and the resulting grape characteristics provide the basis from which winemaking can begin. A great number of varieties are now approved in the European Union as true grapes for wine-growing and viticulture, because of better characteristics for diseases

Vineyard in Brhlovce, Slovakia.

History

The earliest evidence of grape vine cultivation and winemaking dates back 7,000 years. The history of viticulture is closely related to the history of wine, with evidence that humans cultivated wild grapes to make wine as far back as the Neolithic period. Evidence suggests that some of the earliest domestication of *Vitis vinifera* occurred in the area of the modern countries Georgia and Armenia. The oldest-known winery was discovered in the «Areni-1» cave in Vayots Dzor, Armenia. Dated to c. 4100 BC, the site contained a wine press, fermentation vats, jars, and cups. Archaeologists also found *V. vinifera* seeds and vines. Commenting on the importance of the find, McGovern said, "The fact that winemaking was already so well developed in 4000 BC suggests that the technology probably goes back much earlier." There is also evidence of grape domestication in the Near East in the early Bronze Age, around 3200 BC.

Evidence of ancient viticulture is provided by cuneiform sources (ancient writing on clay tablets), plant remains, historical geography, and archaeological excavations. The remnants of ancient wine jars have been used to determine the culture of wine consumption and cultivated grape species. In addition to winemaking, grapes have been grown for the production of raisins. The earliest act of cultivation appears to have been the favoring of hermaphroditic members of the *Vitis vinifera* species over the barren male vines and the female vines, which were dependent on a nearby male for pollination. With the ability to pollinate itself, over time the hermaphroditic vines were able to sire offspring that were consistently hermaphroditic.

The period that Thucydides was most likely referencing was the time between 3000 BC and 2000 BC, when viticulture emerged in force in Asia Minor, Greece, and the Cyclades Islands of the Aegean Sea. During this period, grape cultivation developed from an aspect of local consumption to an important component of international economies and trade.

Roman

From 1200 BC to 900 BC, the Phoenicians developed viticulture practices that were later used in Carthage. Around 500 BC, the Carthaginian writer Mago recorded such practices in a two-volume work that was one of the few artifacts to survive the Roman destruction of Carthage during the Third Punic War. The Roman statesman Cato the Elder was influenced by these texts, and around 160 BC he wrote *De Agricultura,* which expounded on Roman viticulture and agriculture. Around 65 AD, the Roman writer Columella produced the most detailed work on Roman viticulture in his twelve-volume text *De Re Rustica*. Columella's work is one of the earliest to detail trellis systems for raising vines off the ground. Columella advocated the use of stakes versus the previously accepted practice of training vines to grow up along tree trunks. The benefits of using stakes over trees was largely to minimize the dangers associated with climbing trees, which was necessary to prune the dense foliage in order to give the vines sunlight, and later to harvest them.

Roman expansion across Western Europe brought Roman viticulture to the areas that would become some of the world's best-known wine-growing regions: the Spanish Rioja, the German Mosel, and the French Bordeaux, Burgundy and Rhône. Roman viticulturists were among the first to identify steep hillsides as one of the better locations to plant vines, because cool air runs downhill and gathers at the bottom of valleys. While some cool air is beneficial, too much can rob the vine of the heat it needs for photosynthesis, and in winter it increases the risk of frost.

Medieval

In the Middle Ages, Catholic monks (particularly the Cistercians) were the most prominent viticulturists of the time period. Around this time, an early system of *Metayage* emerged in France with laborers (*Prendeur*) working the vineyards under contractual agreements with the landowners (*Bailleur*). In most cases, the *prendeurs* were given flexibility in selecting their crop and developing their own vineyard practice. In northern Europe, the weather and climate posed difficulties for grape cultivation, so certain species were selected that better suited the environment. Most vineyards grew white varieties of grape, which are more resistant to the damp and cold climates. A couple species of red grape, such as the Pinot Noir, were also introduced.

Many of the viticultural practices developed in this time period would become staples of European viticulture till the 18th century. Varietals were studied more intently to see which vines were the most suitable for a particular area. Around this time, an early concept of *terroir* emerged as wines from particular places began to develop a reputation for uniqueness. The concept of pruning for quality over quantity emerged, mainly through Cistercian labors, though it would create conflict between the rich landowners who wanted higher quality wines and the peasant laborers whose livelihood depended on the quantity of wine they could sell. The Riesling is the famous example for higher quality of wine. In 1435 Count John IV. of Katzenelnbogen started this successful tradition.

In Burgundy, the Cistercian monks developed the concept of *cru* vineyards as homogeneous pieces of land that consistently produce wines each vintage that are similar. In areas like the Côte-d›Or the monks divided the land into separate vineyards, many of which are still around today—like Montrachet and La Romanée.

Les Très Riches Heures du duc de Berry dates back to 1416 and depicts horticulture and viticulture in France. The images illustrate peasants bending down to prune grapes from vines behind castle walls. Additional illustrations depict grape vines being harvested, with each vine being cut to three spurs around knee height.

In Mythology

In Greek mythology, the demigod Dionysus (Roman name: Bacchus) – son of Zeus – invented the grapevine and the winepress. When his closest satyr friend died trying to bring him the vine that Dionysus thought was important, Dionysus forced the vine to bear fruit and he later invented wine. His fame spread and he finally became a God.

The Grape Vine

The vast majority of the world's wine producing regions are found between the temperate latitudes of 30° and 50° in each hemisphere. Within these bands, the annual mean temperatures are between 10 and 20 °C (50 and 68 °F). The presence of large bodies of water and mountain ranges can have positive effects on the climate and vines. Nearby lakes and rivers can serve as protection for drastic temperature drops at night by releasing the heat that the water has stored during the day to warm the vines.

The Berry

The grape is classified as a berry. On the vine, grapes are organized through systems known as clusters. Grape clusters also can vary in compactness which can result in long clusters (resulting in the grapes spreading out) or short clusters (resulting in grapes packed together). In some grape species, clusters ripen collectively which allow them to be harvested together. For others, grapes may ripen individually within a cluster. Each grape berry contains a pedicel which attaches to the rachis. The main function of the rachis is to allow the grapes receive their water and nutrients. The pollination and fertilization of grapes results in one to four seeds within the berry. When fertilization does not occur, seedless grapes are formed, which are sought after for the production of raisins. Regardless of pollination and fertilization, most plants will produce around 100 to 200 grapes.

The skin of the grape accounts for 5 to 20% of the total weight of a grape depending on the variety. When grape skin ripens, it contains the majority of the aromatic substances and tannin. These factors become important in winemaking for methods including color extraction or aroma dissolution. Although the skin contains the majority of the tannin, small percentages can be found throughout the grape and during all of its developmental stages. However, the tannin's most important role is during the grape›s ripening stage as its function is to formulate color and body shape.

Growing Vines

Although many factors can affect the overall quality of a grape vine, the three most important are climate, slope, and soil.

Climate

Climate is the most significant factor in determining a grape's inherent qualities. Each grape species has a uniquely preferred environment for ideal growing. Because climates vary from country to country, selecting the best strain is an important decision in grape cultivation. In addition, because climatic factors such as temperature and rain can be unpredictable and uncontrollable, each year will produce unique qualities and yields of grapes.

Grape vines need approximately 1300–1500 hours of sunshine during the growing season and around 690 millimetres (27 in) of rainfall throughout the year in order to produce grapes suitable for winemaking. In ideal circumstances, the vine will receive most of the rainfall during the winter and spring months: rain at harvesttime can create many hazards, such as fungal diseases and berry splitting. The optimum weather during the growing season is a long, warm summer that allows the grapes the opportunity to ripen fully and to develop a balance between the levels of acids and sugars in the grape. Hot and sunny climates have a frost-free growing season of 200 days or more. These climates allow grapes to ripen faster with higher sugar levels and lower acidity. Cooler climates have a frost-free growing season of around 150–160 days. Cooler seasons force the grapes to ripen earlier which produces a fresher and more acidic harvest. In general, the average yearly temperature for most crops should average around 15 °C (59 °F) in order for the highest quality to be achieved in each grape. Summer: Ideal temperatures in summer average around 22 °C (72 °F). Ideal summer temperatures enable fruits to ripen. Temperature and sunshine are the most important factors in ripening.

Winter: Ideal temperatures in winter average around 3 °C (37 °F). Ideal winter temperatures are necessary to allow grape vines to enter their resting phase. If temperatures fall too low, the crops can be injured. Spring and Fall: Spring and fall are critical seasons for grape development, because the plants are susceptible to frost damage, which can injure the fruiting buds. Wet weather in spring can increase the odds of mildew formation. To prevent mildew, some farms introduce devices such as heaters or large fans in vineyards. However, such solutions can be costly.

Slope

Hillsides and slopes are preferred over flatter terrain: vines growing on a slope can receive a greater intensity of the sun's rays, with sunshine falling on an angle perpendicular to the hillside. In flatter terrain, the intensity of the sunlight is diluted as it spreads out across a wider surface area. Small slopes that are elevated above surrounding ground are the best and safest places for crops, because these small elevations are less prone to frost. Additionally, a slope affords better drainage, obviating the possibility that the vine might sit in overly moist soil. In cooler regions of the northern hemisphere, south-facing slopes receive more hours of sunlight and are preferred; in warmer climes, north-facing slopes are preferred. In the southern hemisphere, these orientations are reversed.

Soil

Quality soil is important to allow plants to have better root systems. The growth and health of a vine can be affected if the soil quality is poor. Different grape species prefer various soil conditions, although there are general quality factors. Favorable soil conditions include: aeration, loose texture, good drainage and moderate fertility. Drainage factors are cited as the most important soil characteristic to affect grape vine growth. When root growth is restricted due to bad soil, vine growth and fruit yields lessen and plant survival rates can dip to only a few years.

Hazards

Examples of downy and powdery mildew on a grape leaf.

Manual grape gathering

There are many hazards that a viticulturist needs to be aware of when growing vines. These hazards can have an adverse effect on the wine produced from the grape or kill the vine itself. When the vine is flowering it is very susceptible to weather hazards such as strong winds and hail. Cold temperatures during this period can also bring the onset of *millerandage* which produces clusters with no seeds and varying sizes. Too much heat can have the opposite reaction and produce Coulure that causes grape clusters to either drop to the ground or not fully develop.

Viticultural hazards include:

- Oidium: A powdery mildew which can attack all green parts of the vine. If left untreated, odium can be terminal for the plant. It thrives in cooler temperatures and in the shade. Some North American vine species have evolved to show resistance to the mildew (3:37).
- Downy mildew: Also known as peronospora, thrives in high temperatures and humidity and produces stains on leaves. Peronospora can be treated by spraying plants with copper sulphate. Most American vines are resistant, excluding *Vitis vinifera*.
- Fanleaf: can lead to deformity, yellowing of the leaves, and smaller crop yields. Fanleaf is a virus spread by nematodes which breed in the vine stem. There is no cure for the plant, and the best action to take is to remove the infected plants and allow remaining roots in the soil to rot.
- Frost
- Phylloxera
- Plant virus

FROST

Frost is the coating or deposit of ice that may form in humid air in cold conditions, usually overnight. In temperate climates it most commonly appears as fragile white crystals or frozen dew drops near the ground, but in cold climates it occurs in a greater variety of forms. Frost is composed of delicate branched patterns of ice crystals formed as the result of fractal process development.

Frost is known to damage crops or reduce future crop yields, therefore farmers in those regions where frost is a problem often invest substantial means to prevent its formation. Frost forms when the temperature of a solid surface in the open cools to below the freezing point of water and for the most clearly crystalline forms of frost in particular, below the frost point in still air. In most temperate countries such temperatures usually are the result of heat loss by radiation at night, so those types of frost sometimes are called radiation frost.

Types of frost include crystalline hoar frost from deposition of water vapor from air of low humidity, white frost in humid conditions, window frost on glass surfaces, advection frost from cold wind over cold surfaces, black frost without visible ice at low temperatures and very low humidity, and rime under supercooled wet conditions.

The size of frost crystals varies depending on the time they have been building up and the amount of water vapor available. Frost crystals may be clear or translucent, but, like snow, a mass of frost crystals will scatter light in all directions, so that a coating of frost appears white.

Formation

Frost in the highest town in Venezuela, Apartaderos. Because its location in an alpine tundraecosystem called páramo, they often undergo a sudden and drastic change in a daily freeze-and-thaw cycle, sometimes described as «summer every day and winter every night.»

If a solid surface is chilled below the dew point of the surrounding humid air and the surface itself is colder than freezing, ice will form on it. If the water deposits as a liquid that then freezes, it forms a coating that may look glassy, opaque, or crystalline, depending on its type. Depending on context, that process also may be called atmospheric icing. The ice it produces differs in some ways from crystalline frost, which consists of spicules of ice that typically project from the solid surface on which they grow.

The main difference between the ice coatings and frost spicules arises from the fact that the crystalline spicules grow directly from desublimation of water vapour from air, and desublimation is not a factor in icing of freezing surfaces. For desublimation to proceed the surface must be below the frost point of the air, meaning that it is sufficiently cold for ice to form without passing through the liquid phase. The air must be humid, but not sufficiently humid to permit the condensation of liquid water, or icing will result instead of desublimation. The size of the crystals depends largely on the temperature, the amount of water vapor available, and how long they have been growing undisturbed.

As a rule, except in conditions where supercooled droplets are present in the air, frost will form only if the deposition surface is colder than the surrounding air. For instance frost may be observed around cracks in cold wooden sidewalks when humid air escapes from the warmer ground beneath. Other objects on which frost commonly forms are those with low specific heat or high thermal emissivity, such as blackened metals; hence the accumulation of frost on the heads of rusty nails.

The apparently erratic occurrence of frost in adjacent localities is due partly to differences of elevation, the lower areas becoming colder on calm nights. Where static air settles above an area of ground in the absence of wind, the absorptivity and specific heat of the ground strongly influence the temperature that the trapped air attains.

Types

Hoar Frost

Hoar frost (also hoarfrost, radiation frost, or pruina) refers to white ice crystals, deposited on the ground or loosely attached to exposed objects such as wires or leaves. They form on cold, clear nights when conditions are such that heat radiates out to the open sky faster than it can be replaced from nearby sources such as wind or warm objects. Under suitable circumstances, objects cool to below the frost point of the surrounding air, well below the freezing point of water. Such freezing may be promoted by effects such as flood frost or frost pocket.These occur when ground-level radiation losses cool air till it flows downhill and accumulates in pockets of very cold air in valleys and hollows. Hoar frost may freeze in such low-lying cold air even when the air temperature a few feet above ground is well above freezing.

A spider web covered in air hoar frost

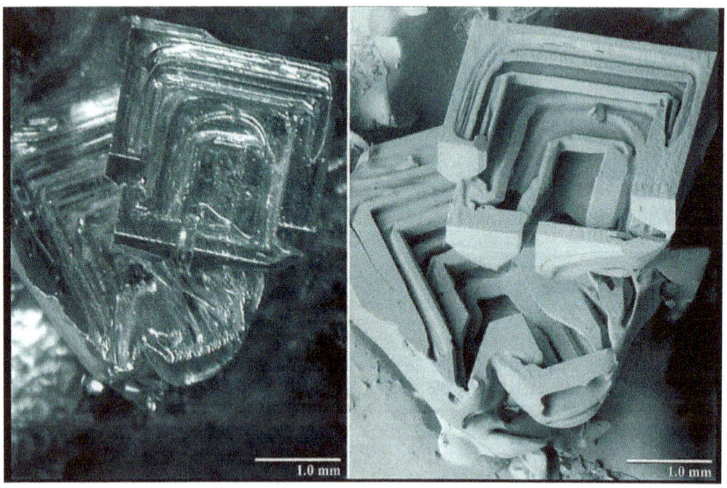

Depth hoar, imaged with light and with scanning electron microscopy.

The name hoar comes from an Old English adjective that means "showing signs of old age"; in this context it refers to the frost that makes trees and bushes look like white hair.

Hoar frost may have different names depending on where it forms:

- **air hoar** is a deposit of hoar frost on objects above the surface, such as tree branches, plant stems, and wires;

- **surface hoar** refers to fern-like ice crystals directly deposited on snow, ice or already frozen surfaces;
- **crevasse hoar** consists of crystals that form in glacial crevasses where water vapour can accumulate under calm weather conditions;
- **depth hoar** refers to faceted crystals that have slowly grown large within cavities beneath the surface of banks of dry snow. Depth hoar crystals grow continuously at the expense of neighbouring smaller crystals, so typically are visibly stepped and have faceted hollows.

When surface hoar covers sloping snowbanks, the layer of frost crystals may create an avalanche risk; when heavy layers of new snow cover the frosty surface, furry crystals standing out from the old snow hold off the falling flakes, forming a layer of voids that prevent the new snow layers from bonding strongly to the old snow beneath. Ideal conditions for hoarfrost to form on snow are cold clear nights, with very light, cold air currents conveying humidity at the right rate for growth of frost crystals. Wind that is too strong or warm destroys the furry crystals, and thereby may permit a stronger bond between the old and new snow layers. However, if the winds are strong enough and cold enough to lay the crystals flat and dry, carpeting the snow with cold, loose crystals without removing or destroying them or letting them warm up and become sticky, the frost interface between the snow layers still may present an avalanche danger, because the texture of the frost crystals differs from the snow texture and the dry crystals will not stick to fresh snow. Such conditions still prevent a strong bond between the snow layers.

In very low temperatures where fluffy surface hoar crystals form without subsequently being covered with snow, strong winds may break them off, forming a dust of ice particles and blowing them over the surface. The ice dust then may form yukimarimo, as has been observed in parts of Antarctica, in a process similar to the formation of dust bunnies and similar structures.

Hoar frost and white frost also occurs in man-made environments such as in freezers or industrial cold storage facilities. If such cold spaces or the pipes serving them are not well insulated and are exposed to ambient humidity, the moisture will freeze instantly depending on the freezertemperature. The frost may coat pipes thickly, partly insulating them, but such inefficient insulation still is a source of heat loss.

Advection Frost

Advection frost (also called wind frost) refers to tiny ice spikes that form when there is a very cold wind blowing over branches of trees, poles and other surfaces. It looks like rimming on the edge of flowers and leaves and usually it forms against the direction of the wind. It can occur at any hour, day or night.

A flower with advection frost on the tips of its petals.

Window Frost

Window frost (also called fern frost or ice flowers) forms when a glass pane is exposed to very cold air on the outside and warmer, moderately moist air on the inside. If the pane is not a good insulator (for example, if it is a single pane window), water vapour condenses on the glass forming frost patterns. With very low temperatures outside, frost can appear on the bottom of the window even with double pane energy efficient windows because the air convection between two panes of glass ensures that the bottom part of the glazing unit is colder than the top part. On unheated motor vehicles the frost will usually form on the outside surface of the glass first. The glass surface influences the shape of crystals, so imperfections, scratches, or dust can modify the way ice nucleates. The patterns in window frost form a fractal with a fractal dimension greater than one but less than two. This is a consequence of the nucleation process being constrained to unfold in two dimensions, unlike a snowflake which is shaped by a similar process but forms in three dimensions and has a fractal dimension greater than two.

If the indoor air is very humid, rather than moderately so, water will first condense in small droplets and then freeze into clear ice.

Similar patterns of freezing may occur on other smooth vertical surfaces, but they seldom are as obvious or spectacular as on clear glass.

White Frost

White frost is a solid deposition of ice which forms directly from water vapour contained in air.

White frost forms when there is a relative humidity above 90% and a temperature below −8 °C (18 °F) and it grows against the wind direction, since air arriving from windward has a higher humidity than leeward air, but the wind must not be strong or it damages the delicate icy structures as they begin to form. White frost resembles a heavy coating of hoar frost with big, interlocking crystals, usually needle-shaped.

Rime

Rime is a type of ice deposition that occurs quickly, often under conditions of heavily saturated air and windy conditions. Technically speaking, it is not a type of frost, since usually supercooled water drops are involved, in contrast to the formation of hoar frost, in which water vapour desublimates slowly and directly. Ships travelling through Arctic seas may accumulate large quantities of rime on the rigging. Unlike hoar frost, which has a feathery appearance, rime generally has an icy solid appearance.

Black Frost

Black frost (or «killing frost») is not strictly speaking frost at all, because it is the condition seen in crops when the humidity is too low for frost to form, but the temperature falls so low that plant tissues freeze and die, becoming blackened, hence the term «black frost». Black frost often is called «killing frost» because white frost tends to be less cold, partly because the latent heat of freezing of the water reduces the temperature drop.

Effect on Plants

Overview

Frost on the grass of a public park in January.

Many plants can be damaged or killed by freezing temperatures or frost. This varies with the type of plant, the tissue exposed, and how low temperatures get: a "light frost" of −2 to 0 °C (28 to 32 °F) will damage fewer types of plants than a "hard frost" below −2 °C (28 °F). Plants likely to be damaged even by a light frost include vines—such as beans, grapes, squashes, melons—along with nightshades such as tomatoes, eggplants and peppers. Plants that may tolerate (or even benefit) from frosts include:

- root vegetables (e.g. beets, carrots, parsnips, onions)
- leafy greens (e.g. lettuces, spinach, chard, cucumber)
- cruciferous vegetables (e.g. cabbages, cauliflower, bok choy, broccoli, Brussels sprouts, radishes, kale, collard, mustard, turnips, rutabagas)

Even those plants that tolerate frost may be damaged once temperatures drop even lower (below −4 °C or 25 °F). Hardy perennials, such as *Hosta*, become dormant after the first frosts and regrow when spring arrives. The entire visible plant may turn completely brown until the spring warmth, or may drop all of its leaves and flowers, leaving the stem and stalk only. Evergreen plants, such as pine trees, withstand frost although all or most growth stops. Frost crack is a bark defect caused by a combination of low temperatures and heat from the winter sun.

Vegetation is not necessarily damaged when leaf temperatures drop below the freezing point of their cell contents. In the absence of a site nucleating the formation of ice crystals, the leaves remain in a supercooled liquid state, safely reaching temperatures of −4 to −12 °C (25 to 10 °F). However, once frost forms, the leaf cells may be damaged by sharp ice crystals. Hardening is the process by which a plant becomes tolerant to low temperatures. See also cryobiology.

Certain bacteria, notably *Pseudomonas syringae*, are particularly effective at triggering frost formation, raising the nucleation temperature to about −2 °C (28 °F). Bacteria lacking ice nucleation-active proteins (ice-minus bacteria) result in greatly reduced frost damage.

Protection Methods

Curitiba (Southern Brazil) is the coldest of Brazil›s state capitals; the greenhouse of the Botanical Garden of Curitiba protects sensitive plants.

Typical measures to prevent frost or reduce its severity include one or more of:

- deploying powerful blowers to simulate wind, thereby preventing the formation of accumulations of cold air. There are variations on this theme. One of them is the selective inverted sink prevents frost by drawing cold air from the ground and blowing it up through a chimney. It was originally developed to prevent frost damage to citrus fruits in Uruguay. In New Zealand, helicopters are used in a similar function, especially in the vineyard regions like Marlborough. By dragging down warmer air from the inversion layers, and preventing the ponding of colder air on the ground, the low-flying helicopters prevent damage to the fruit buds. As the operations are conducted at night, and have in the past involved up to 130 aircraft per night in one region, safety rules are strict. Although not a dedicated method, wind turbines have similar (small) effect of vertically mixing air layers of different temperature.

- for high-value crops, farmers may wrap trees and cover crops.
- heating to slow the drop in temperature. This is not practical except for high value crops grown over small areas.
- production of smoke to reduce cooling by radiation
- spraying crops with a layer of water that releases latent heat, preventing harmful freezing of the tissues of the plants that it coats.

Such measures need to be applied with discretion, because they may do more harm than good; for example, spraying crops with water can cause damage if the plants become overburdened with ice. An effective low cost method for small crop farms and plant nurseries, exploits the latent heat of freezing. A pulsed irrigation timer delivers water through existing overhead sprinklers at a low volumes to combat frosts down to −5 °C (23 °F). If the water freezes it giving off its latent heat, preventing the temperature of the foliage from falling much below zero.

Frost-Free Areas

Frost-free areas are found mainly in the tropics, where they cover almost all land except at altitudes above about 3,000 metres or 9,800 feet near the equator and around 2,000 metres or 6,600 feet in the semi-arid middle tropics, but also in areas with subtropical climates that have winters tempered by strong oceanic influences. The most poleward frost-free areas are the lower altitudes of the Azores, Île Amsterdam, Île Saint-Paul, and Tristan da Cunha.

The only reliably frost-free areas in the contiguous United States are the Florida Keys and the coastal areas of the Channel Islands of California.

Personifications

Further information: Ded Moroz and Father Frost (fairy tale)

Frost is personified in Russian culture as Ded Moroz. Indigenous peoples of Russia such as the Mordvins have their own traditions of frost deities.

English folklore tradition holds that Jack Frost, an elfish creature, is responsible for feathery patterns of frost found on windows on cold mornings.

Green Harvest

A green harvest is the removal of immature grape bunches, typically for the purpose of increasing yield. The removal of the bunches while they are still green induces the vine to put all its energy into developing the remaining grapes. In theory this results in better ripening and the development of more numerous and mature flavour compounds. In the absence of a green harvest, a healthy, vigorous vine can produce dilute, unripe grapes.

In Europe, many appellations restrict the yield permitted from a given area, so there is even more incentive to perform green harvesting when presented with excess crop. Often the excess must be sold for a pittance and used for industrial alcohol production rather than wine.

While the concept of thinning or sacrificing part of the grape crop, i.e. green harvesting, with the aim of improving the quality of the remaining grapes, predates modern critics, the practice has increased in recent times in vineyards found in California and areas where the grapes grow easily. (McCoy)

Field Blend

Mechanical harvesting of Sauvignon blanc grapes in Côtes de Duras France

A field blend is a wine that is produced from two or more different grape varieties interplanted in the same vineyard. In the days before precise varietal identification, let alone rigorous clonal selection, a vineyard might be planted by taking cuttings from another vineyard and therefore approximately copying its genetic makeup. This meant that one vine could be Zinfandel and the next Carignan. When making wine with little equipment to spare for separate vinification of different varieties, field blends allowed effortless, though inflexible, blending.

Fermentation tanks are now cheap enough that the field blend is an anachronism, and almost all wines are assembled by blending from smaller, individual lots. However, in California some of the oldest (and lowest-yielding) Zinfandel comes from vineyards that are field-blended. Ridge Vineyards owns the Lytton Springs vineyards in Sonoma County, which were planted from 1900 to 1905 in what Ridge calls "a traditional field blend of about seventy percent Zinfandel, twenty percent Petite Sirah, and ten percent Grenache and Carignan."

Gemischter Satz (*Mixed set*) is a wine term in German equivalent to a field blend, which means that grapes of different varieties are planted, harvested and vinified together. In older times, this was common, but the practice has almost stopped. It is, however, a specialty of Vienna.

REFERENCES

1. Johnson, H. (1989). Vintage: The Story of Wine. Simon & Schuster. pp. 17–19. ISBN 0-671-68702-6.
2. Philips, Roderick (2000). A Short History of Wine. New York: Ecco.
3. "Books: Roots of the Vine". Archaeology.org. Retrieved 2012-02-12.
4. David Keys (28 December 2003). "Now that's what you call a real vintage: professor unearths 8,000-year-old wine". The Independent. independent.co.uk. Retrieved 13 January 2011.
5. Mark Berkowitz (September–October 1996). "World's Earliest Wine". Archaeology. Archaeological Institute of America. 49 (5). Retrieved 13 January 2011.
6. "'Oldest known wine-making facility' found in Armenia". BBC News. BBC. 11 January 2011. Retrieved 13 January 2011.
7. Thomas H. Maugh II (11 January 2011). "Ancient winery found in Armenia". Los Angeles Times. Los Angeles Times Media Group. Retrieved 13 January 2011.
8. McGovern, Patrick (1996). The Origins and Ancient History of Wine. Amsterdam, The Netherlands: Gordon and Breach. p. 99.
9. McGovern, Patrick (1996). The Origins and Ancient History of Wine. Amsterdam, The Netherlands: Gordon and Breach. p. 198.
10. McGovern, Patrick (1996). The Origins and Ancient History of Wine. Amsterdam, The Netherlands: Gordon and Breach. p. 104.
11. Johnson, H. (1989). Vintage: The Story of Wine. Simon & Schuster. p. 18. ISBN 0-671-68702-6.
12. Johnson, H. (1989). Vintage: The Story of Wine. Simon & Schuster. p. 35. ISBN 0-671-68702-6.
13. Johnson, H. (1989). Vintage: The Story of Wine. Simon & Schuster. pp. 35–39. ISBN 0-671-68702-6.
14. Johnson, H. (1989). Vintage: The Story of Wine. Simon & Schuster. p. 61. ISBN 0-671-68702-6.
15. Johnson, H. (1989). Vintage: The Story of Wine. Simon & Schuster. p. 68. ISBN 0-671-68702-6.
16. Johnson, H. (1989). Vintage: The Story of Wine. Simon & Schuster. pp. 82–92. ISBN 0-671-68702-6.

17. Johnson, H. (1989). Vintage: The Story of Wine. Simon & Schuster. p. 116. ISBN 0-671-68702-6.
18. Unwin, Tim (1991). Wine and the Vine: An Historical Geography of Viticulture and the Wine Trade. London: Routledge. p. 169.
19. Unwin, Tim (1991). Wine and the Vine: An Historical Geography of Viticulture and the Wine Trade. London: Routledge. p. 171.
20. Johnson, H. (1989). Vintage: The Story of Wine. Simon & Schuster. pp. 121–122. ISBN 0-671-68702-6.
21. "Katzenelnbogener Weltrekorde: Erster RIESLING und erste BRATWURST!". Graf-von-katzenelnbogen.de. Retrieved 2012-02-12.
22. Johnson, H. (1989). Vintage: The Story of Wine. Simon & Schuster. pp. 131–132. ISBN 0-671-68702-6.
23. Fraga, H., Garcia de C. A. I., Malheiro, A.C., Santos, J.A., 2016. Modelling climate change impacts on viticultural yield, phenology and stress conditions in Europe. Global Change Biology: doi:10.1111/gcb.13382.
24. Ribéreau-Gayon, P (2000). Handbook of Enology Vol. 1. Chichester, West Sussex, England: John Wiley. p. 220.
25. Vine, Richard (1997). Winemaking: From Grape Growing to Marketplace. New York: Chapman & Hall. p. 24. ISBN 9780412122217.
26. Ribéreau-Gayon, P (2000). Handbook of Enology Vol. 1. Chichester, West Sussex, England: John Wiley. p. 221.
27. Vine, Richard (1997). Winemaking: From Grape Growing to Marketplace. New York: Chapman & Hall. p. 223. ISBN 9780412122217.
28. Unwin, Tim (1991). Wine and the Vine: An Historical Geography of Viticulture and the Wine Trade. London: Routledge. p. 36.
29. Vine, Richard (1997). Winemaking: From Grape Growing to Marketplace. New York: Chapman & Hall. p. 224. ISBN 9780412122217.
30. Unwin, Tim (1991). Wine and the Vine: An Historical Geography of Viticulture and the Wine Trade. London: Routledge. p. 42.
31. Stevenson, T. (2005). The Sotheby's Wine Encyclopedia. pp. 14–15. ISBN 0-7566-1324-8.
32. Vine, Richard (1997). Winemaking: From Grape Growing to Marketplace. New York: Chapman & Hall. p. 26. ISBN 9780412122217.
33. Stevenson, T. (2005). The Sotheby's Wine Encyclopedia. p. 16. ISBN 0-7566-1324-8.
34. Vine, Richard (1997). Winemaking: From Grape Growing to Marketplace. New York: Chapman & Hall. p. 27. ISBN 9780412122217.
35. Unwin, Tim (1991). Wine and the Vine: An Historical Geography of Viticulture and the Wine Trade. London: Routledge. p. 44.
36. Unwin, Tim (1991). Wine and the Vine: An Historical Geography of Viticulture and the Wine Trade. London: Routledge. p. 37.

37. Unwin, Tim (1991). Wine and the Vine: An Historical Geography of Viticulture and the Wine Trade. London: Routledge. p. 38.
38. Unwin, Tim (1991). Wine and the Vine: An Historical Geography of Viticulture and the Wine Trade. London: Routledge. p. 39.
39. "Frost – Definition of frost by Merriam-Webster". merriam-webster.com.
40. John E. Oliver (1 January 2005). The Encyclopedia of World Climatology. Springer Science & Business Media. pp. 382–. ISBN 978-1-4020-3264-6.
41. "What causes frost?". Retrieved 2007-12-05.
42. "Hoarfrost – Definition of hoarfrost by Merriam-Webster". merriam-webster.com.
43. Louisiana State University Baton Rouge Robert V Rohli; Robert V. Rohli; Anthony J. Vega (13 December 2013). Climatology. Jones & Bartlett Publishers. pp. 381–. ISBN 978-1-284-05427-9.
44. "Weather Facts: Frost hollow – Weather UK – weatheronline.co.uk". weatheronline.co.uk.
45. David McClung; Peter A. Schaerer (2006). The Avalanche Handbook. The Mountaineers Books. pp. 72–. ISBN 978-0-89886-809-8.
46. West, Bruce; Mauro Bologna (2003). Physics of Fractal Operators. Paolo Grigolini. Springer. p. 46. ISBN 0-387-95554-2.
47. "Rime – Definition of rime by Merriam-Webster". merriam-webster.com.
48. "Frost Tolerance of Vegetables". Botanical Interests. Retrieved Nov 12, 2013.
49. "Fall vegetables vs. Summer vegetables".
50. Klosinska, Urszula; et al. (27 February 2014), "Low temperature seed germination of cucumber: genetic basis of the tolerance trait", Journal of Horticultural Research, de Gruyter, 21 (2), doi:10.2478/johr-2013-0031, retrieved 2014-02-28
51. Maki LR, Galyan EL, Chang-Chien MM, Caldwell DR (1974). "Ice Nucleation Induced by Pseudomonas syringae". Applied Microbiology. 28 (3): 456–459. PMC 186742 . PMID 4371331.
52. Lindow, Stephen E.; Deane C. Arny; Christen D. Upper (October 1982). "Bacterial Ice Nucleation: A Factor in Frost Injury to Plants". Plant Physiology. 70 (4): 1084–1089. doi:10.1104/pp.70.4.1084. PMC 1065830 . PMID 16662618.
53. Selective Inverted Sink Rolex Awards site (won award in Technology and Innovation category) 1998.
54. Helicopters Fight Frost – Vector, Civil Aviation Authority of New Zealand, September / October 2008, Page 8-9
55. Turbines and turbulence, Nature (journal), 468, 1001, 23 December 2010, DOI:10.1038/4681001a, published online 22 December 2010.
56. Somnath Baidya Roy and Justin J. Traiteur. Impacts of wind farms on surface air temperatures, Proceedings of the National Academy of Sciences, Vol. 107, No. 42, October 19, 2010, p. 17,899.
57. Wind farms impacting weather, Science Daily.

58. "A practical method of frost protection". Retrieved 31 October 2011.
59. Selders, Arthur W. "Frost protection with sprinkler irrigation"(PDF). West Virginia university. Retrieved 31 October 2011.

CHAPTER 8

Tropical Gardening

Chapter Contents

INTRODUCTION

Tropical horticulture is a branch of horticulture that studies and cultivates plants in the tropics, i.e., the equatorial regions of the world. The field is sometimes known by the portmanteau "TropHort".

Tropical horticulture includes plants such as perennial woody plants (arboriculture), ornamentals (floriculture), vegetables (olericulture), and fruits (pomology) including grapes (viticulture). The origin of many of these crops is not in the tropics but in temperate zones. Their adoption to tropical climatic conditions is an objective of breeding. Many important crops, however, are indigenous to the tropics. The latter embrace perennial crops such as oil palm, vegetables including okra, field crops such as rice and sugarcane, and particularly fruits including pineapple, banana, papaya, and mango.

Since the tropics represent 36 percent of the earth›s surface and 20 percent of its land surface, the potential of tropical horticulture is huge. In contrast to temperate regions, environmental conditions in the tropics are defined less by seasonal temperature fluctuations and more by seasonality of precipitation. Thus the climate in the greater part of the tropics is characterized by distinct wet and dry seasons, although such variation is reduced in locations closer to the equator (±5° latitude). Temperature conditions in the tropics are affected by elevation, in which contrasting warmer and colder climate areas in the tropics can be differentiated,[4-5] and highland areas in the tropics can consequently be more favourable for production of temperate plant species than are lowland areas.

Latex being collected from an incised rubber tree (*Hevea brasiliensis*), and a bucket of collected latex

Types of Plants

Both vascular and non-vascular plants grow in tropical environments. Plants indigenous to the tropics are usually cold sensitive and adapted to receiving high levels of solar radiation. They are sensitive to small variations in photoperiod («short day» plants), and can be adapted to extended drought, high precipitation and/or distinct wet and dry seasons. High night temperatures are a major hindrance to adopting temperate crops (e.g., tomatoes) to the tropical lowlands. Furthermore, such conditions promote high respiration rates of plants, resulting in comparably lower net photosynthesis rates.

TROPICAL GARDEN

A tropical garden features tropical plants and requires good rainfall or a decent irrigation or sprinkler system for watering. These gardens typically need fertilizer and heavy mulching.

Tropical gardens are no longer exclusive to tropical areas. Many gardeners in colder climates are adopting the tropical garden design, which is possible through careful choice of plants and flowers. Main features include plants with very large leaves, vegetation that builds in height towards the back of the garden, creating a dense garden. Large plants and small trees hang over the garden, leaving sunlight to hit the ground directly.

A tropical garden on the Portuguese island of Madeira.

Tropical plants

The following are some examples of tropical plants to be used in tropical gardens or as indoor plants.

Hoyas

Hoyas are known as the "Wax Flowers" due to their texture and almost unreal appearance. They need relatively warm and humid conditions, but depending on the specific variety of Hoya, they can endure heavy rainfall for some months during the wet seasons just as they may also be exposed to long dry periods. They are commonly used indoors. They need indirect light, but the amount or required light depends on the variety. They need small amounts of water and the soil is to be kept slightly dry since a very wet soil can end up killing this kind of plant. They can be grown in pots that are not too big or in hanging baskets. They grow better if kept at a minimum temperature of 15 degree Celsius (59 Fahrenheit).

Banana Plants

Contrary to common belief, growing banana trees is not that difficult and allow people to enjoy their own bananas. Also these plants can be used as windbreaks. They need fertile soils, large mulch and organic matter, large amounts of nitrogen and potassium, warm temperature, high humidity, large amounts of water, and shelter from other banana plants. Banana plants are not to be exposed to strong winds and extreme weather conditions (too hot or too cold weather) with an ideal temperature being 26-30 degree Celsius (78-86 Fahrenheit). They stop growing below 14 degree Celsius (57 Fahrenheit). Banana rhizomes are planted upright and their roots have to be well covered with soil.

Bird of Paradise

Among the many species of birds of paradise is the Strelitzia reginae, an indigenous plant to South Africa and South America. As a tropical plant, it grows in warm, humid climates. This exotic, colorful plant, with evergreen leaves resembles a bird's peak. It is an outdoor plant provided the weather is not too cold. Otherwise, it is better to keep it in a pot indoors. It needs rich soil as well as sun or partial shade. Another requirement for this plant is a good drainage.

Ferns

Ferns are commonly used to give gardens a great foliage. Most ferns are easy to take care of. They basically cannot be exposed to direct sunlight for long periods of time. Moreover, their soil should always be moist. One way to keep ferns moist is by misting them.

Papaya Tree

Papaya trees should be planted where they can be kept warm and free from wind and freeze. They need enough water to support their leaves, but it should not be exceed as to a root rod. They also need a well-drained soil.

Orchids

Night temperatures between 12- 18 degree Celsius (55-65 Fahrenheit), and day temperatures between 23 -26 degree Celsius (75-80 Fahrenheit) are fine temperatures for orchids. Orchids need sufficient amounts of light as well as humidity. However, there are many varieties of orchids and each of them features specific care needs.

Maintenance

A tropical garden is one of the most difficult gardens to build, or maintain, it becomes more difficult the more your local climate differs from the natural habitat of your plants. Key to a healthy tropical garden are lots of light and lots of water. The large leaves that feature in tropical plants require the soil to be humid at all times, so irrigation might be a must-have for some gardens. Over-watering can kill your plants as well, as it will cause the roots to rot.

A tropical plant that is not cold-hardy should be brought indoors during the winter and returned to the garden for the summer.

Non-Tropical Climates

A greenhouse is a good way to grow tropical plants in a colder climate.

Tropical plants that work well in non-tropical climates include;

- acanthus mollis
- Algerian ivy
- canna
- crocosmia, aka L
- paulownia tomentosa
- secrecia
- trachycarpus

Acanthus Mollis

Acanthus mollis, commonly known as bear's breeches, sea dock, bearsfoot or oyster plant, is a herbaceous perennial plant with an underground rhizome in the genus Acanthus. It is regarded as an invasive species in some jurisdictions.

Etymology

The name of the genus derives from the Greek term κανθος, akanthos, the name of the plant in Greek; it was imitated in Corinthian capitals; the word is related to κανθα, *akantha* meaning "thorn" referring to the thorn-bearing sepals, or any thorny or prickly plant in Greek, while the Latin name of the species, *mollis* meaning "soft", "smooth", refers to the texture of the leaves.

Description

Close-up on a flower of *Acanthus mollis*

Leaf of *Acanthus mollis*

Corinthian column capital in the Mosque of Uqba in Kairouan, Tunisia

Acanthus mollis reaches on average 30–80 centimetres (12–31 in) of height, with a maximum of 180 centimetres (71 in), inflorescence included. It has basal clusters of deeply lobed and cut, shiny dark green leaves, soft to the touch, up to 40 centimetres (16 in) long and 25 centimetres (9.8 in) broad, with a long petiole. The inflorescence is a cylindrical spike 30–40 centimetres (12–16 in) long and can produce up to 120 flowers. The flowers are tubular, whitish, and lilac or rose in colour. Each flower is up to 5 centimetres (2.0 in) long and it is surrounded by three green or purplish bracts. The central bract is spiny and larger than the other two. The calyx has two lips: the upper is purple on top, rather long and forms a kind of «helmet» on top of the corolla. The corolla is reduced to a white lower lip, trilobed, with purple-pink venation. The four stamens are fused to the corolla and look like tiny brushes. This species flowers in late spring or early summer, from May through August.

Reproduction

Acanthus mollis is entomophilous and it is pollinated only by bees or bumble bees large enough to force their way between the upper sepal and the lower, so that they can reach the nectar at the bottom of the tube. The fruit is an ovoid capsule containing two to four large black seeds. The dispersal of seeds is by the wind (anemochory).

These plants are usually propagated from tubers and tend to form large, localized clumps which can survive for several decades and look statuesque when well-grown, but its suitability as a garden plant is lessened on account of its invasive nature (new plants are produced readily both from seed and portions of root) and its susceptibility to powdery mildew and attacks from snails.

Distribution

This plant is native to the Mediterranean region from Portugal and northwest Africa east to Croatia and it is one of the earliest cultivated species.

Habitat

This garden plant is also quite common in the wild. It grows in dry areas, roadsides and wastelands, especially in rocky and bushy places. It is tolerant of drought and shade and generally does not exceed an altitude of 300 metres (980 ft) above sea level.

Culture

The leaves of this plant are generally considered by historians to have been the design inspiration for the Corinthian order capitals of Greco-Roman architecture. Virgil describes Helen of Troy as wearing a dress embroidered with acanthus leaves.

Gallery

Plants of Acanthus mollis

Inflorescence of Acanthus mollis

Close-up on stamens of Acanthus mollis

Acanthus mollis - MHNT

HEDERA ALGERIENSIS

Hedera algeriensis (Algerian ivy; syn. Hedera canariensis var. algeriensis Willd) is a species of evergreen ivy native to the North African coast, including coastal mountains in Algeria.

Description

It is a vigorous, large evergreen climber, self-clinging by aerial roots. The stems are reddish, partly decorated with stellate hairs or scales, reddish-starred with about 15 rays. The leaves are alternate and simple. As in other *Hedera* species, the fertile branches have stiff stems with leaves having usually fewer lobes compared to the sterile (juvenile) stems. The main part of the leaf (the lamina) is oval-rhombic, 12-20 cm by 5–12 cm on flowering branches and coarsely toothed or slightly lobed (3-5 lobes) on the sterile stems. The inflorescence is a cluster of 13-15 pubescent flowers. Small flowers in umbels develop only on the fertile stems or branches. The plant contains the glycoside hederagenin, mostly in the leaves and berries, which could cause a mild toxicosis. If these parts are ingested it may cause severe discomfort, and handling plants may cause skin irritation or an allergic reaction. The plant is rejected by herbivorous animals. Variegated cultivars of *Hedera algeriensis* are widely used in gardening. It naturalizes in mild climates and lowlands.

The cultivar ‹Ravensholst› has gained the Royal Horticultural Society›s Award of Garden Merit.

Ecology

It may be a noxious weed or be invasive. It requires consistently moist soil, but is more resistant to environmental dryness than most related species and can stand drought better than other ivies. It tolerates low temperatures better than island species of similar latitudes, such as *Hedera canariensis*, *Hedera maderensis*, *Hedera azorica*, or *Hedera cypria*. It is native to the North African coast and coastal mountains where the climate is cooler and condensation provides a more humid habitat. It is frost sensitive at about the -2 °C winter isotherm but can become naturalised in mild climates. *Hedera algeriensis* arose from a common ancestor in the cloud forest habitat in the Mediterranean area. The northern African *Hedera* and northeast Atlantic species of the genus *Hedera* are closely related species. Until recently it was thought there was a single species, *Hedera helix*, but recent studies have shown that there are several species that differ mainly by microscopic details of the hairiness of the buds. *Hedera maroccana*, *Hedera iberica*, and *Hedera canariensis* are closely related.

Landscaping

In Coastal California Algerian ivy is used as a ground cover on highway embankments to help control erosion. Without irrigation it can be difficult to establish and maintain, and can hide underlying dead, dry material, which can be a fire hazard. The ivy groundcover provides cover, food and herbage for roof rats and snails.

CANNA (PLANT)

Canna (or **canna lily**, although not a true lily) is a genus of 10 species of flowering plants. The closest living relations to cannas are the other plant families of the order Zingiberales, that is the Zingiberaceae (gingers), Musaceae (bananas), Marantaceae, Heliconiaceae, Strelitziaceae, etc.

Canna is the only genus in the family Cannaceae. The APG II system of 2003 also recognizes the family, and assigns it to the order Zingiberales in the clade commelinids, in the monocots.

The species have large, attractive foliage, and horticulturists have turned it into a large-flowered and bright garden plant. In addition, it is one of the world›s richest starch sources, and is an agricultural plant.

Although a plant of the tropics, most cultivars have been developed in temperate climates and are easy to grow in most countries of the world as long as they receive at least 6–8 hours average sunlight during the summer, and are moved to a warm location for the winter. See the *Canna* cultivar gallery for photographs of *Canna* cultivars.

The name *Canna* originates from the Latin word for a cane or reed.

Botany

The plants are large tropical and subtropical perennial herbs with a rhizomatous rootstock.

The broad, flat, alternate leaves that are such a feature of this plant, grow out of a stem in a long, narrow roll and then unfurl. The leaves are typically solid green, but some cultivars have glaucose, brownish, maroon, or even variegated leaves.

Canna (Foliage Group) 'Auguste Ferrier'

The flowers are composed of three sepals and three petals that are seldom noticed by people, they are small and hidden under extravagant stamens. What appear to be petals are the highly modified stamens or staminodes. The staminodes number (1–) 3 (–4) (with at least one staminodal member called the labellum, always being present. A specialized staminode, the stamen, bears pollen from a half-anther. A somewhat narrower ‹petal› is the pistil which is connected down to a three-chambered ovary.

The flowers are typically red, orange, or yellow or any combination of those colours, and are aggregated in inflorescences that are spikes or panicles (thyrses). Although gardeners enjoy these odd flowers, nature really intended them to attract pollinators collecting nectar and pollen, such as bees, hummingbirds, sunbirds, and bats. The pollination mechanism is conspicuously specialized. Pollen is shed on the style while still in the bud, and in the species and early hybrids some is also found on the stigma because of the high position of the anther, which means that they are self-pollinating. Later cultivars have a lower anther, and rely on pollinators alighting on the labellum and touching first the terminal stigma, and then the pollen.

The wild species often grow to at least 2–3 m (6.6–9.8 ft) in height, but there is a wide variation in size among cultivated plants; numerous cultivars have been selected for smaller stature. Cannas grow from swollen underground stems, correctly known as rhizomes, which store starch, and this is the main attraction of the plant to agriculture, having the largest starch particles of all plant life.

Canna is the only member of the Liliopsida class (monocot group) in which hibernation of seed is known to occur, due to its hard, impenetrable seed covering.

Distribution

Canna 'Pink Dancer' in Escambray Mountains, Cuba

The genus is native to tropical and subtropical regions of the New World, from the southern United States (southern South Carolina west to southern Texas) and south to northern Argentina.

Canna indica has become naturalized in many tropical areas around the world, is a difficult plant to remove, and is invasive in some places.

Canna cultivars are grown in most countries, even those with territory above the Arctic Circle, which have short summers but long days, and the rapid growth rate of cannas makes them a feasible gardening plant, as long as they receive 6–8 hours of sunlight each day during the growing season and are protected from the cold of winter.

Origins

Canna (Crozy Group) 'Alberich', Pfitzer 1949

Canna indica, commonly called achira in Latin America, has been cultivated by Native Americans in tropical America for thousands of years and was one of the earliest domesticated plants in the Americas. The starchy root is edible.

The first species of *Canna* introduced to Europe was *C. indica* L., which was imported from the East Indies, though the species originated from the Americas. Charles de l›Ecluse, who first described and sketched *C. indica*, indicated this origin, and stated that it was given the name *indica*, not because the plant is from India, in Asia, but because this species was originally transported from America: *Quia ex America primum delata sit*; and at that time, one described the tropical areas of that part of the globe as the Western Indies.

Much later, in 1658, Pison made reference to another species which he documented under the vulgar or common name of 'Albara' and 'Pacivira', which resided, he said, in the 'shaded and damp places, between the tropics'; this species is *Canna angustifolia* L. (later reclassified as *C. glauca* L. by taxonomists).

Without exception, all *Canna* species that have been introduced into Europe can be traced back to the Americas, and it can be asserted with confidence that *Canna* is solely an American genus. If Asia and Africa provided some of the early introductions, they were only varieties resulting from *C. indica* and *C. glauca* cultivars that have been grown for a long time in India and Africa, with both species imported from Central and South America. Since cannas have very hard and durable seed coverings, it is likely that seed remains would survive in the right conditions and been found by archaeologists in the Old World if *Canna* had been grown there from antiquity. If the soils of India or Africa had produced some of them, they would have been imported before the 1860s into European gardens.

Uses

Canna (Italian Group) 'Yellow King Humbert'

Detail of the seed pods and seeds: The seeds are used for jewelry and musical instruments.

- Some species and many cultivars are widely grown in the garden in temperate and subtropical regions. Sometimes, they are also grown as potted plants. A large number of ornamental cultivars have been developed. They can be used in herbaceous borders, tropical plantings, and as a patio or decking plant.
- Internationally, cannas are one of the most popular garden plants and a large horticultural industry depends on the plant.
- The rhizomes of cannas are rich in starch, and it has many uses in agriculture. All of the plant has commercial value, rhizomes for starch (consumption by humans and livestock), stems and foliage for animal fodder, young shoots as a vegetable, and young seeds as an addition to tortillas.
- The seeds are used as beads in jewelry.
- The seeds are used as the mobile elements of the kayamb, a musical instrument from Réunion, as well as the hosho, a gourd rattle from Zimbabwe, where the seeds are known as *hota* seeds.
- In more remote regions of India, cannas are fermented to produce alcohol.
- The plant yields a fibre from the stem, which is used as a jute substitute.
- A fibre obtained from the leaves is used for making paper. The leaves are harvested in late summer after the plant has flowered, they are scraped to remove the outer skin, and are then soaked in water for two hours prior to cooking. The fibres are cooked for 24 hours with lye and then beaten in a blender. They make a light tan brown paper.
- A purple dye is obtained from the seed.
- Smoke from the burning leaves is said to be insecticidal.

- Cannas are used to extract many undesirable pollutants in a wetland environment as they have a high tolerance to contaminants.
- In Thailand, cannas are a traditional gift for Father›s Day.
- In Vietnam, canna starch is used to make cellophane noodles known as *mi n dong*.

Taxonomy

Species

Although most cannas grown these days are cultivars (see below), there are approximately 20 known species of the wild form, and in the last three decades of the 20th century, *Canna* species have been categorized by two different taxonomists, Paul Maas, from the Netherlands and Nobuyuki Tanaka from Japan. Both reduced the number of species from the 50-100 accepted previously, assigning most as synonyms.

This reduction in species is also confirmed by work done by Kress and Prince at the Smithsonian Institution; however, this only covers a subset of the species range.

See List of *Canna* species for full species information and descriptions.

Horticultural Varieties (Cultivars)

Canna 'Golden Gate'

Canna (Crozy Group) 'Theresa Blakey'

See the List of Canna cultivars for photographs of Canna cultivars.

Cannas became very popular in Victorian times as garden plants and were grown widely in France, Germany, Hungary, India, Italy, the United Kingdom, and the United States. Some cultivars from this time, including a sterile hybrid, usually referred to as *Canna* × *ehemannii*, are still commercially available. *C.* × *ehemannii* is tall and green leafed with terminal drooping panicles of hot pink iris like flowers, looking somewhat like a cross between a banana and a fuchsia.

As tender perennials in northern climates, they suffered severe setbacks when two world wars sent the young gardening staff off to war. It took many years for the frugalities of war and its rationing subsequences to change to the more prosperous times of the late 20th century. The *Canna* genus has recently experienced a renewed interest and revival in popularity.

Once, hundreds of cultivars existed, but many are now extinct. In 1910, Árpäd Mühle, from Hungary, published his *Canna* book, written in German. It contained descriptions of over 500 cultivars.

In recent years, many new cultivars have been created, but the genus suffers severely from having many synonyms for many popular ones. Most of the synonyms were created by old varieties resurfacing without viable names, with the increase in popularity from the 1960s onwards. Research has accumulated over 2,800 *Canna* cultivar names, but many of these are simply synonyms.

See List of *Canna* hybridists for details of the people and firms that created the current *Canna* legacy.

In the early 20th century, Professor Liberty Hyde Bailey defined, in detail, two 'garden species' (*C.* × *generalis* and *C.* × *orchiodes*) to categorise the floriferous cannas being grown at that time, namely the Crozy hybrids and the 'orchid-like' hybrids introduced by Carl Ludwig Sprenger in Italy and Luther Burbank in the USA, at about the same time (1894). The definition was based on the genotype, rather than the phenotype, of the two cultivar groups. Inevitably, over time those two floriferous groups were interbred, the distinctions became blurred and overlapped, and the Bailey species names became redundant. Pseudo-species names are now deprecated by the International Code of Nomenclature for Cultivated Plants which, instead, provides Cultivar Groups for categorising cultivars (see groups at List of *Canna* cultivars).

Agricultural Varieties

Canna (Agriculture Group) 'Edulis Dark'

The Canna Agriculture Group contains all of the varieties of *Canna* grown in agriculture. "Canna achira" is a generic term used in South America to describe the cannas that have been selectively bred for agricultural purposes, normally derived from *C. discolor*. It is grown especially for its edible rootstock from which starch is obtained, but the leaves and young seeds are also edible, and achira was once a staple food crop in Peru and Ecuador. Trials in Ecuador using a wide range of varieties have shown that achira can yield on average 56 tons of rhizomes and 7.8 tons of extractable starch per hectare. However, the crop needs 9–12 months to mature to full productivity.

Many more traditional kinds exist worldwide; they have all involved human selection, so are classified as agricultural cultivars. Traditionally, *Canna edulis* Ker Gawl. has been reputed to be the species grown for food in South America, but *C. edulis* probably is simply a synonym of *C. discolor*, which is also grown for agricultural purposes throughout Asia.

Cultivation

Canna (Miniature Group) 'Oberon'

Cannas grow best in full sun with moderate water in well-drained rich or sandy soil. They grow from perennial rhizomes, but are frequently grown as annuals in temperate zones for an exotic or tropical look in the garden. In arid regions, cannas are often grown in the water garden, with the lower inch of pot submerged. In all areas, high winds tear the leaves, so shelter is advised.

The rhizomes are frost tender and will rot if left unprotected in freezing conditions. In areas which go below about −10 °C (14 °F) in the winter (< USDA Zone 8b), the rhizomes can be dug up before freezing and stored in a protected area (above 7 °C or 45 °F) for replanting in the spring. Otherwise, they should be protected by a thick layer of mulch over winter.

These canna cultivars have gained the Royal Horticultural Society›s Award of Garden

Pests

Cannas are largely free of pests, but in the US, plants sometimes fall victim to the canna leaf roller moth, and the resultant leaf damage, while not fatal to the plant, can be most distressing to a keen gardener.

Slugs and snails are fond of cannas and can leave large holes in the leaves, preferring the tender young unfurled leaves. Red spider mites can also be a problem for cannas grown indoors or during a very hot, long summer outdoors. Japanese beetles can also ravage the leaves if left uncontrolled.

Disease

Cannas are remarkably free of diseases, compared to many genera. However, they may fall victim to canna rust, a fungus resulting in orange spots on the plant's leaves, caused by over moist soil. They are also susceptible to certain plant viruses, some of which are *Canna*-specific, which may result in spotted or streaked leaves, in a mild form, but can finally result in stunted growth and twisted and distorted blooms and foliage.

The flowers are sometimes affected by a grey, fuzzy mold called botrytis. Under humid conditions, it is often found growing on the older flowers. Treatment is to simply remove the old flowers, so the mold does not spread to the new flowers.

Propagation

Sexual Propagation

Seeds are produced from sexual reproduction, involving the transfer of pollen from the stamen of the pollen parent onto the stigma of the seed parent. In the case of *Canna*, the same plant can usually play the roles of both pollen and seed parents, technically referred to as a hermaphrodite. However, the cultivars of the Italian Group and triploids are almost always seed sterile, and their pollen has a low fertility level. Mutations are almost always totally sterile.

Canna fruit (green) and ripe seed pods

Canna fruits

Canna seeds have a very hard seed coat, which contributes to their dormancy. Germination is facilitated by scarification of the seed coat, which can be accomplished by several techniques.

Pollination

The species are capable of self-pollination, but most cultivars require an outside pollinator. All cannas produce nectar, so attract nectar-consuming insects, bats, and hummingbirds, that act as the transfer agent, spreading pollen between stamens and stigmas on the same or different flowers.

Genetic Changes

Since genetic recombination has occurred, a cultivar grown from seed will have different characteristics from its parent(s), thus should never be given a parent›s name. The wild species have evolved in the absence of other *Canna* genes and are usually true to type when the parents are of the same species, but a degree of variance still occurs. The species *C. indica* is an aggregate species, having many different and extreme forms ranging from the giant to miniature, from large foliage to small foliage, both green and dark foliage, and many different coloured blooms, red, orange, pink, and yellow, and combinations of those colours.

Asexual Propagation

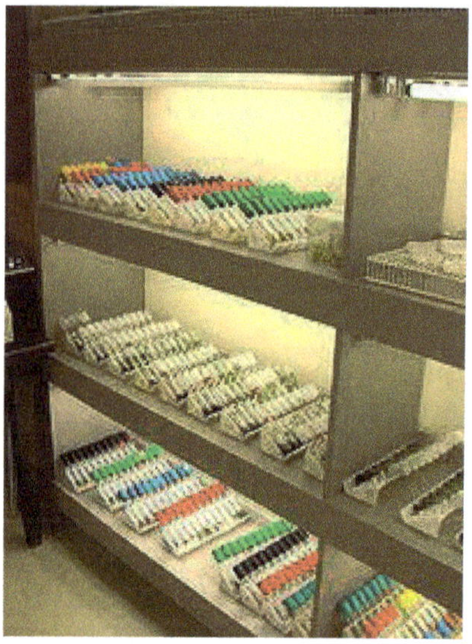

In vitro culture of plants in a controlled, sterile environment

Division of Plant Parts

Outside of a laboratory, the only effective asexual propagation method is rhizome division. This uses material from a single parent, and as there is no exchange of genetic material, it almost always produces plants that are identical to the parent. After a summer's growth, the horticultural cultivars can be separated into typically four or five separate smaller rhizomes, each with a growing nodal point ('growing eye'). Without the growing point, which is composed of meristem material, the rhizome will not grow.

Micropropagation

Micropropagation, also known as tissue culture, is the practice of rapidly multiplying stock plant material to produce a large number of progeny plants. Micropropagation uses *in vitro* division of small pieces in a sterile environment, where they first produce proliferations of tissue, which are then separated into small pieces that are treated differently so that they produce roots and new stem tissue. The steps in the process are regulated by different ratios of plant growth regulators. Many commercial organizations have produced cannas this way, and specifically the "Island Series" of cannas was introduced by means of mass-produced plants using this technique. However, cannas have a reputation for being difficult micropropagation candidates.

Micropropagation techniques can be employed to disinfest plants of a virus. In the growing tip of a plant, cell division is so rapid that the younger cells may not have had time to be infected with the virus. The rapidly growing region of meristem cells producing the shoot tip is cut off and placed *in vitro*, with a very high probability of being uncontaminated by virus.

PAULOWNIA TOMENTOSA

Paulownia tomentosa (common names princesstree, foxglove-tree, empress tree or kiri) is a deciduous tree in the family Paulowniaceae, native to central and western China. Paulownia tomentosa is an extremely fast-growing tree, and is a persistent exotic invasive in North America.

Nomenclature

Flowering tree

The generic name *Paulownia* honors Anna Pavlovna of Russia. The specific epithet *tomentosa* is a Latin word meaning 'covered in hairs'.

Description

Flowers And Young Leaves

This tree grows 10–25 m (33–82 ft) tall, with large heart-shaped to five-lobed leaves 15–40 cm (6–16 in) across, arranged in opposite pairs on the stem. On young growth, the leaves may be in whorls of three and be much bigger than the leaves on more mature growth. The leaves can be mistaken for those of the catalpa.

The very fragrant flowers are produced before the leaves in early spring, on panicles 10–30 cm long, with a tubular purple corolla 4–6 cm long resembling a foxglove flower. The fruit is a dry egg-shaped capsule 3–4 cm long, containing numerous tiny seeds. The seeds are winged and disperse by wind and water. Pollarded trees do not produce flowers, as these only form on mature wood.

Paulownia tomentosa can survive wildfire because the roots can regenerate new, very fast-growing stems. It is tolerant of pollution and it is not fussy about soil type. For this reason it functions ecologically as a pioneer plant. Its nitrogen-rich leaves provide good fodder and its roots prevent soil erosion. Eventually, *Paulownia* is succeeded by taller trees that shade it. It cannot thrive in the shade of other trees.

Uses

Twig with old seed pods

Paulownia tomentosa is cultivated as an ornamental tree in parks and gardens. It has gained the Royal Horticultural Society›s Award of Garden Merit.

The characteristic large size of the young growth is exploited by gardeners: by pollarding the tree and ensuring there is vigorous new growth every year, massive leaves are produced (up to 23in/60 cm across). These are popular in the modern style of gardening which uses large-foliaged and «architectural» plants.

In China, the tree is planted at the birth of a girl. The fast-growing tree matures when she does. When she is eligible for marriage the tree is cut down and carved into wooden articles for her dowry. Carving the wood of *Paulownia* is an art form in Japan and China. In legend, it is said that the phoenix will only land on the Empress Tree and only when a good ruler is in power. Several Asian string instruments are made from *P. tomentosa*, including the Japanese *koto* and Korean *gayageum* zithers.

The soft, lightweight seeds were commonly used as a packing material by Chinese porcelain exporters in the 19th century, before the development of polystyrene packaging. Packing cases would often leak or burst open in transit and scatter the seeds along rail tracks. The magnitude of the numbers of seeds used for packaging, together with seeds deliberately planted for ornament, has allowed the species to be viewed as an invasive species in areas where the climate is suitable for its growth, notably Japan and the eastern United States.

Composition

Some geranylflavonoids can be found in *P. tomentosa*.

Verbascoside can also be produced in hairy roots cultures of *P. tomentosa*.

Flower close-up

Trunk

Fruit

Spent Fruit

Leaf

TRACHYCARPUS

Trachycarpus is a genus of nine species of palms native to Asia, from the Himalaya east to eastern China. They are fan palms (Arecaceae tribe Corypheae), with the leaves with a bare petiole terminating in a rounded fan of numerous leaflets. The leaf bases produce persistent fibres that often give the trunk a characteristic hairy appearance. All species are dioecious, with male and female flowers produced on separate plants although female plants will sometimes produce male flowers, allowing occasional self-pollination.

Cultivation and Uses

Stem of *Trachycarpus fortunei* showing leaf base fibres

The most common species in cultivation is *Trachycarpus fortunei* (Chusan palm or windmill palm), which is, in cultivated range, the northernmost palm species in the world. There are several hardy *Trachycarpus* species that stay outdoors all year long with little to no protection at the New York Botanical Garden. The dwarf form popularly known as *T. wagnerianus* is unknown in the wild, and is now considered synonymous with *T. fortunei*. It resembles that species closely, differing only in its smaller and stiffer leaves. Hybrids between them are intermediate in size and fully fertile.

Trachycarpus takil (the Kumaon palm) is similar to *T. fortunei*; and probably even hardier. Other species less common in cultivation are *T. geminisectus*, *T. princeps*, *T. latisectus*, *T. martianus*, *T. nanus* and *T. oreophilus*. *T. martianus* and *T. latisectus* do not tolerate cold as well as *T. fortunei* or *T. takil*. *T. geminisectus*, *T. princeps* and *T. oreophilus* are still too rare and small in cultivation to assess their full potential.

The trunk fibres produced by the leaf sheaths of *Trachycarpus fortunei* are harvested in China and elsewhere to make coarse but very strong rope, brooms and brushes.This use gives rise to the old alternative name «hemp-palm». The fibrous leaf sheaths are also frequently used to clothe the stems of artificial palms.

This genus is very popular among palm enthusiasts for its ability to withstand cold, especially in the form of damp, cool summer weather with relatively mild winter weather. These palms often tolerate snow in their native habitats and are the hardiest trunking palms.

Trachycarpus species are used as food plants by the larvae of some Lepidoptera species including *Paysandisia archon* (recorded on *T. fortunei*).

CULINARY FRUITS

This list of culinary fruits contains the names of some fruits that are considered edible in some cuisines. The word "fruit" is used in several different ways. The definition of fruit for these lists is a culinary fruit, i.e. "Any sweet, edible part of a plant that resembles fruit, even if it does not develop from a floral ovary; also used in a technically imprecise sense for some sweet or sweetish vegetables, some of which may resemble a true fruit or are used in cookery as if they were a fruit, for example rhubarb." Note that many edible plant parts that are true fruits botanically speaking, are not considered culinary fruits. They are classified as vegetables in the culinary sense (for example: the tomato, zucchini, and so on), and hence they do not appear in this list. Similarly, some botanical fruits are classified as nuts (e.g. Brazil nut and various almonds), or staples (e.g. breadfruit), and likewise do not appear here. There also exist many fruits which are edible and palatable but for various reasons have not become popular.

Some culinary fruits

Alphabetical List by Common Name

Papayas

An array of tropical fruits at University of Hyderabad, India

Acerola

Mangos

Passion fruits

A

- Abiu (*Pouteria caimito*; Sapotaceae)
- Açaí (*Euterpe oleracea*; Arecaceae), or Assai
- Acerola (*Malpighia glabra*; Malpighiaceae), also called Barbados cherry or West Indian cherry
- Ackee (*Blighia sapida* or *Cupania sapida*; Sapindaceae)
- African cherry orange (*Citropsis articulata*; Rutaceae)
- African mango (*Irvingia gabonensis*)
- African moringa (*Moringa stenopetala*; Moringaceae)
- Amazon Tree-Grape Moraceae(*Pourouma cecropiaefolia*)
- Ambarella (*Spondias dulcis*; Anacardiaceae)
- American black elderberry (*Sambucus canadensis*; Adoxaceae)
- American chestnut (*Castanea dentata*; Fagaceae)
- American grape: North American species (e.g., *Vitis labrusca*; Vitaceae) and American-European hybrids are grown where grape (*Vitis vinifera*) is not hardy and are used as rootstocks
- American hazelnut (*Corylus americana*; Betulaceae)
- American pawpaw (*Asimina triloba*; Annonaceae)
- American persimmon (*Diospyros virginiana*; Ebenaceae)
- American plum (*Prunus americana*; Rosaceae)

- American red elderberry (*Sambucus pubens*; Adoxaceae)
- American red raspberry (*Rubus strigosus*; Rosaceae)
- Annona
- Apple and crabapple (*Malus*)
- Apple, Malay, see Malay apple
- Apple, custard, see custard apple
- Apple, elephant, see elephant apple
- Apple, kei, see key apple
- Apple, mammee, see mammee apple
- Apricot (*Prunus armeniaca* or *Armeniaca vulgaris*)
- Araza (*Eugenia stipitata*)
- Arhat (*Siraitia grosvenorii*; Cucurbitaceae) Also called longevity fruit
- Atemoya (*Annona cherimola* × *Annona squamosa*; Annonaceae)
- Atherton Raspberry (*Rubus probus*; Rosaceae)

B

- Babaco (*Carica pentagona*; Caricaceae)
- Bacupari (*Garcinia gardneriana*)
- Bacuri (*Platonia esculenta*; Guttiferae)
- Bael, or woodapple (*Aegle marmelos*; Rutaceae), found in eastern India.
- Banana (*Musacea* spp.; Musaceae); its starchy variant is the plantain
- Barbadine (granadilla; maracujá-açu in Portuguese)
- Barbados cherry, see acerola
- Berberis vulgaris
- Batuan (*Garcinia morella*)
- Beach plum (*Prunus maritima*; Rosaceae)
- Bearberry (*Arctostaphylos* spp.)
- Betel nut ("Areca catechu"; Arecaceae)
- Bignay (*Antidesma bunius*; Euphorbiaceae)
- Bilberry or whortleberry (*Vaccinium* spp.)
- Bilimbi (*Averrhoa bilimbi*; Oxalidaceae) Also called cucumber tree or tree sorrel
- Biriba (*Rollinia deliciosa*; Annonaceae)
- Bitter melon, the flesh of which is bitter, and used as a culinary vegetable, but with a sweet coating around the mature seeds
- Black apple (*Planchonella australis*; Sapotaceae)
- Black cherry (*Prunus serotina*; Rosaceae) very popular flavoring for pies, jams, and sweets.

- Blackcurrant ("Ribes nigrum")
- Black raspberry (*Rubus occidentalis* or *Rubus leucodermis*; Rosaceae)
- Black mulberry (*Morus nigra*; Moraceae)
- Black sapote (*Diospyros digyna*)
- Black walnut (*Juglans nigra*; Juglandaceae)
- Blackberry, of which there are many species and hybrids, such as dewberry, boysenberry, olallieberry, and tayberry (genus *Rubus*)
- Blood orange
- Blue tongue (*Melastoma affine*; Melastomataceae)
- Blueberry (*Vaccinium, sect. Cyanococcus*; Ericaceae)
- Bolivian coconut (*Parajubaea torallyi*)
- Bolwarra (*Eupomatia laurina*; Eupomatiaceae)
- Boquila (*Boquila trifoliata* ; Lardizabalaceae)
- Bottle gourd also known as calabash (*Lagenaria siceraria*; Cucurbitaceae)
- Rubus ursinus × idaeus *; (*Rosaceae*)*
- Brazilian guava (*Psidium guineense*; Myrtaceae)
- breadfruit ("Artocarpus altilis"; Artocarpus)
- Breadnut (*Artocarpus camansi*; Moraceae)
- Breadnut, Mayan, see Mayan breadnut
- Broad-leaf bramble (*Rubus moluccanus*; Rosaceae)
- Brazilian guava (*Psidium guineense*; Myrtaceae)
- Brush cherry (*Syzygium australe*; Myrtaceae)
- Buddha's hand
- Buffaloberry ("Shepherdia argentea"; Elaeagnaceae), which grows wild in the prairies of Canada
- Burdekin plum (*Pleiogynium timoriense*; Anacardiaceae)
- Burmese grape, or latka (*Baccaurea sapida*; Phyllanthaceae)
- Bush tomato (Certain *Solanum* species; Solanaceae)
- Button mangosteen (*Garcinia prainiana*)

C

- Caimito (*Chrysophyllum_cainito*; Sapotaceae)
- Calabash tree (*Crescentia cujete*)
- Calamondin (*Citrofortunella Microcarpa*)
- Calamansi (×*Citrofortunella Mitis*)
- Camu camu (*Myrciaria dubia*; Myrtaceae)
- Candlenut fruit (*Aleurites moluccana*)

- Canary melon
- Canistel (*Pouteria campechiana*; Sapotaceae), also called yellow sapote or "eggfruit»
- Cantaloupe
- Cape gooseberry (*Physalis peruviana*; Solanaceae)
- Capuli cherry (*Prunus salicifolia*, *Prunus capuli* or *Prunus serotina subsp. capuli*); Rosaceae), native to the Andes
- Carambola (*Averrhoa carambola*; Oxalidaceae), also called star fruit or *five fingers*
- Cardón (*Pachycereus pringlei*; Cactaceae)
- Carob (*Ceratonia siliqua*; Fabaceae)
- Cashew apple (*Anacardium occidentale*)
- Cassabanana (*Sicana odorifera*; Cucurbitaceae)
- Cattley guava (*Psidium cattleianum*; Myrtaceae)
- Cawesh (*Annona scleroderma*; Annonaceae)
- Cedar Bay cherry (*Eugenia carissoides*; Myrtaceae)
- Ceriman (*Monstera deliciosa*; Araceae)
- Cereus peruvianus
- Ceylon gooseberry (*Dovyalis hebecarpa*; Flacourtiaceae)
- Charichuelo (*Garcinia intermedia*)
- Chalta (*Dillenia indica*)
- Champedak (*Artocarpus integer*)
- Chayote (*Sechium edule*; Cucurbitaceae)
- Che (*Cudrania tricuspidata*; Moraceae) Also called cudrania, Chinese mulberry, cudrang, Mandarin melon berry, silkworm thorn, zhe
- Chenet (guinep or ackee; pitomba-das-Guinas in Portuguese), also known as Spanish lime or mamoncillo
- Cherimoya (*Annona cherimola*; Annonaceae), native to the Ecuadorian Andes
- Cherry, sweet, black, sour, and wild species (*Prunus avium*, *Prunus serotina*, *Prunus cerasus*, and others)
- Cherry ballart (*Exocarpus cupressiformis*; Santalaceae)
- Cherry of the Rio Grande (*Eugenia aggregata*; Myrtaceae)
- Chilean guava (*Psidium cattleianum*; see also ugni (Myrtaceae)
- Chinese jujube (*Ziziphus zizyphus*; Rhamnaceae)
- Chinese olive (*Canarium album*; Burseraceae)
- Chinese quince (*Pseudocydonia sinensis*)
- Chokeberry (*Aronia*)
- Chokecherry (*Prunus virginiana*)

- Chupa-chupa (*Quararibea cordata*; Malvaceae)
- Citron (*Citrus medica*)
- Clementine (*Citrus reticulata* var. *clementine*)
- Cloudberry (*Rubus chamaemorus*)
- Cluster fig (*Ficus racemosa*; Moraceae)
- Cocky apple (*Planchonia careya*)
- Coco plum (*Chrysobalanus icaco*; Chrysobalanaceae)
- Cocona (*Solanum sessilifolium*; Solanaceae)
- Coconut (*Cocos nucifera*; Arecaceae)
- Cocoplum (*Chrysobalanus icaco*; Chrysobalanaceae)
- Coffee ("coffee cherries" surround the better-known "bean")
- Cola nut (*Cola acuminata*; Sterculiaceae)
- Common apple-berry (*Billardiera scandens*; Pittosporaceae)
- Conkerberry (*Carissa lanceolata*; Apocynaceae)
- Cornelian cherry (*Cornus mas*; Cornaceae)
- Costa Rican guava (*Psidium friedrichsthalianum*; Myrtaceae)
- Cranberry (*Vaccinium* spp.)
- Crowberry (*Empetrum* spp.)
- Cupuaçu (*Theobroma grandiflorum*; Malvaceae)
- Currant (*Ribes* spp.; Grossulariaceae), red, black, and white types
- Curry-leaf tree (*Murraya koenigii*; Rutaceae)
- Custard apple (*Annona reticulata*; Annonaceae), also called *bullock's heart*

D

- Dabai (*Canarium odontophyllum*; Burseraceae)
- Damson plum (*Prunus domestica subsp. insititia*; Rosaceae)
- Date (*Phoenix dactylifera*; Arecaceae)
- Date plum (*Diospyros lotus*; Ebenaceae)
- Davidson's plum (*Davidsonia* spp.; Cunoniaceae) Davidsonia jerseyana, Davidsonia johnsonii, Davidsonia pruriens
- Dead man›s fingers (*blue bean, blue sausage fruit, Decaisnea fargesii*)
- Desert banana (*Marsdenia australis*)
- Desert fig (*Ficus platypoda*; Moraceae)
- Desert lime (*Citrus glauca*; Rutaceae)
- Dodder laurel (*Cassytha melantha*)
- Doubah (*Marsdenia australis*; Apocynaceae)
- Double coconut or coco de mer(*Lodoicea maldivica*; Arecaceae)

- Dragonfruit / pitaya (*Hylocereus* spp.; Cactaceae)
- Duku (*Lansium domesticum*; Meliaceae)
- Durian (*Durio zibethicus*; Malvaceae)

E

- Eastern may hawthorn (*Crataegus aestivalis*; Rosaceae, better known as mayhaw.)
- Elderberry (*Sambucus*; Caprifoliaceae)
- Elephant apple (*Dillenia indica*; Dilleniaceae or *Limonia acidissima*; Rutaceae)
- Emblica, see Indian gooseberry
- Emu apple (*Owenia acidula*; Meliaceae)
- Emu berry (*Grewia retusifolia*)

F

- Falberry (*Vaccinium* spp.)
- False-mastic (*Sideroxylon foetidissimum*; Sapotaceae)
- Feijoa, see pineapple guava
- Fibrous satinash (*Syzygium fibrosum*; Myrtaceae)
- Fig (*Ficus* spp. Moraceae)
- Finger lime (*Citrus australasica*; Rutaceae)
- Florida strangler fig (*Ficus aurea*; Moraceae)
- Forest strawberries, Fragaria vesca

G

- Gac
- Galia melon
- Gambooge (*Garcinia cambogia*; Clusiaceae)
- Genip (*Melicoccus bijugatus*; Sapindaceae)
- Giant Colombian blackberry (*Rubus macrocarpus*), native to the highlands of Colombia, up to 5 cm long and 2.5 cm wide
- Giant granadilla (*Passiflora quadrangularis*; Passifloraceae)
- Gooseberry (*Ribes* spp.; Grossulariaceae)
- Goji berries («Lycium barbarum»)
- Gooseberry, Ceylon, see Ceylon gooseberry
- Gooseberry, Indian, see Indian gooseberry
- Gooseberry, Otaheite, see Otaheite gooseberry
- Gooseberry, cape, see Cape gooseberry
- Goumi (*Elaeagnus multiflora*; Elaeocarpaceae)
- Governor's plum (*Flacourtia indica*; Flacourtiaceae)

- Granadilla, Montessa, see Montessa granadilla
- Granadilla, yellow, see yellow granadilla
- Grape, called raisin, sultana when it is dried. (*Vitis* spp.; Vitaceae)
- Grapefruit (*Citrus* × *paradisi*; Rutaceae)
- Greengage, a cultivar of the plum
- Ground plum (*Astragalus crassicarpus*; Fabaceae), also called ground-plum milk-vetch
- Grumichama (*Eugenia brasiliensis*; Myrtaceae)
- Guanabana (*Annona muricata*; Annonaceae)
- Guarana (*Paullinia cupana*; Sapindaceae)
- Guava (*Psidium guajava*; Myrtaceae)
- Guava, Cattley, see Cattley guava
- Guava, Chilean, see Chilean guava
- Guava, Costa Rican, see Costa Rican guava
- Guava, pineapple, see pineapple guava
- Guava, purple, see purple guava
- Guava, strawberry, see strawberry guava
- Guavaberry/Rumberry; (*Myrciaria floribunda*; Myrtaceae)

H

- Hackberry (*Celtis* spp.; Cannabaceae)
- Hairless rambutan
- Hardy kiwi (*Actinidia arguta*; Actinidiaceae family)
- Hawthorn (*Crataegus* and *Rhaphiolepis*)
- Hog plum (taperebá in Portuguese)
- Honeydew
- Edible honeysuckle, haskap, honeyberry
- Horned melon (*Cucumis metuliferus*; Cucurbitaceae)
- Huckleberry (*Vaccinium* spp.)
- Huito (*Genipa americana*; Rubiaceae); also called *jagua*, *genipap*, jenipapo

I

- Ice cream bean (*Inga edulis*)
- Illawarra plum (*Podocarpus elatus*; Podocarpaceae)
- Imbe (*Garcinia livingstonei*)
- Indian almond
- Indian gooseberry (*Phyllanthus emblica/Emblica officinalis*; Euphorbiaceae)
- Indian jujube (*Ziziphus mauritiana*; Rhamnaceae)

- Indian prune (*Flacourtia rukam*; Flacourtiaceae)

J

- Jabuticaba (*Myrciaria cauliflora*; Myrtaceae), also called Brazilian grape tree
- Jackfruit (*Artocarpus heterophyllus* Moraceae), also called *nangka*
- Jambul (*Syzygium cumini*; Myrtaceae)
- Japanese persimmon, or Sharon fruit (*Diospyros kaki*; Ebenaceae)
- Jatobá (*Hymenae coubaril*; Leguminosae) Caesalpinioideae)
- Jelly palm (*Butia capitata*)
- Jocote, also called Jamaica plum
- Jujube (*Ziziphus zizyphus*; Rhamnaceae)
- Jujube, Chinese, see Chinese jujube
- Jujube, Indiana, see Indian jujube
- Juniper berry (*Juniperus* spp.), used for flavoring and in making gin

K

- Kabosu (*Citrus Sphaerocarpa*) Rutaceae
- Kaffir lime (*Citrus hystix*)
- Kahikatea (*Dacrycarpus dacrydioides*)
- Kakadu lime (*Citrus gracilis*; Rutaceae)
- Kakadu plum (*Terminalia ferdinandiana*; Combretaceae)
- Kandis (*Garcinia forbesii*; Clusiaceae)
- Kapok (*Ceiba pentandra*; Bombacaceae)
- Karkalla (*Carpobrotus rossii*; Aizoaceae)
- Karonda (*Carissa carandas*; Apocynaceae)
- Kei apple (*Dovyalis caffra*; Flacourtiaceae)
- Kepel fruit (*Stelechocarpus burahol*; Annonaceae)
- Keule (*Gomortega keule*; Gomortegaceae)
- Key lime (*Citrus aurantifolia*)
- Kitembilla (*Dovyalis hebecarpa*; Flacourtiaceae)
- Kwano, see horned melon
- Kiwifruit (*Actinida* spp.; Actinidiaceae)
- Korean melon (*Cucumis melo var. makuwa*)
- Korlan
- Kumquat (*Fortunella* spp.)
- Kumquat, meiwa, see meiwa kumquat
- Kumquat, nagami, see nagami kumquat

- Kundong (*Garcinia* sp.; Clusiaceae)
- Kutjera (*Solanum centrale*; Solanaceae)
- Kwai muk (*Artocarpus hypargyraea*; Moraceae)

L

- Lady apple (*Syzygium suborbiculare*; Myrtaceae)
- Lakoocha (*Artocarpus lakoocha*; Moraceae)
- Langsat (*Lansium domesticum*), also called longkong or duku
- Lanzones (*Lansium domesticum*; Meliaceae)
- Lapsi (*Choerospondias axillaris Roxb.* Anacardiaceae)
- Lardizabala (*Lardizabala biternata*; Lardizabalaceae)
- Lemon (*Citrus limon*)
- Lemon aspen (*Acronychia acidula*; Rutaceae)
- Leucaena
- Lillypilly (*Acmena* spp., *Syzygium* spp.) Used raw and in jam
- Little gooseberry tree (*Buchanania arborescens*; Anacardiaceae)
- Lime
- Limeberry (*Triphasia trifolia*; Rutaceae)
- Limequat (*Citrus aurantifolia* × *Fortunella* spp.; Rutaceae)
- Lingonberry (*Vaccinium vitis-idaea*)
- Loganberry (*Rubus loganobaccus*)
- Longan (*Dimocarpus longan* or *Euphoria longana*; Sapindaceae)
- Loquat (*Eriobotrya japonica*; Rosaceae)
- Louvi (*Flacourtia inermis*; Flacourtaceae)
- Lúcuma (*Pouteria campechiana*; Sapotaceae)
- Lychee (*Litchi chinensis*; Sapindaceae)

M

- Ma-praang (*Bouea macrophylla*; Anacardiaceae)
- Mabolo (*Diospyros discolor*; Ebenaceae) also known as a velvet persimmon
- Macadamia (*Macadamia integrifolia*; Proteaceae)
- Macadamia, rough shell, see rough shell macadamia
- Madrono (*Rheedia acuminata*; Guttiferae)
- Malabar plum (*Syzygium jambos*; Myrtaceae)
- Malay apple (*Syzygium malaccense*; Myrtaceae)
- Mamey sapote (*Pouteria sapota*; Sapotaceae)
- Mammee apple (*Mammea americana*; Guttiferae)

- Mamoncillo (*Melicoccus bijugatus*; Sapindaceae), also known as quenepa, genip or Fijian longan
- Mandarin (*Citrus reticulata*)
- Mangaba (*Hancornia speciosa*; Apocynaceae)
- Mango (*Mangifera indica*; Anacardiaceae)
- Mangosteen (*Garcinia mangostana*; Guttiferae)
- Manila tamarind (*Pithecellobium dulce*; Leguminosae)
- Manoao (*Manoao colensoi*)
- Maqui (*Aristotelia chilensis*; Elaeocarpaceae)
- Marang
- Marula (*Sclerocarya birrea*)
- Breadnut, Mayan (*Brosimum alicastrum*; Moraceae)
- Mayapple (*Podophyllum* spp.; Berberidaceae)
- Mayhaw, see eastern may hawthorn
- Maypop (*Passiflora incarnata*; Passifloraceae)
- Medlar (*Mespilus germanica*)
- Meiwa kumquat (*Fortunella japonica*; Rutaceae)
- Melinjo
- Melon pear
- Midyim (*Austromyrtus dulcis*; Myrtaceae)
- Miracle fruit (*Synsepalum dulcificum*; Sapotaceae)
- Mock strawberry or Indian strawberry (*Potentilla indica*; Rosaceae)
- Mombin, purple, see purple mombin
- Mombin, red, see red mombin
- Mombin, yellow, see yellow mombin
- Monkey jackfruit (*Artocarpus rigidus*; Moraceae)
- Monkey tamarind (*Inga edulis*; Leguminosae)
- Monstera (*Monstera deliciosa*; Araceae) also called Swiss Cheese Plant, Split-leaf Philodendron
- Montessa granadilla (*Passiflora platyloba*; Passifloraceae)
- Mora común (*Rubus adenotrichus*), most common native berry from Mexico to Ecuador
- Mora de Castilla (*Rubus glaucus*), a blackberry native to South and Central America
- Morinda
- Morinda citrifolia
- Mortiño, or Andean blueberry (*Vaccinium floribundum*), undomesticated, abundant

in the northern Andes

- Mountain pepper (*Tasmannia* spp.; Winteraceae)
- Mountain soursop (*Annona montana*; Annonaceae)
- Mulberry (*Morus* spp.; Moraceae) including black, red and white mulberry
- Munydjudj, see wild plum
- Mundu (*Garcinia dulcis*)
- Muntries (*Kunzea pomifera*; Myrtaceae)
- Muscadine (*Vitis rotundifolia*; Vitaceae)
- Muskmelon

N

- Naartjie (*Citrus reticulata*, *Citrus nobilis*)
- Nagami kumquat (*Fortunella margarita*; Rutaceae)
- Nageia (*Nageia* spp.)
- Nance (*Byrsonima crassifolia*; Malpighiaceae)
- Nannyberry or sheepberry (*Viburnum* spp.; Caprifoliaceae)
- Naranjilla (*Solanum quitoense*; Solanaceae)
- Native currant (*Acrotriche depressa*; Ericaceae)
- Native gooseberry (*Physalis minima*; Solanaceae)
- Native raspberry (Rubus parviflorus)
- Nectarine, see peach
- Neem (*Azadirachta indica*; Meliaceae)
- Néré (*Parkia biglobosa*)
- Nonda plum (*Parinari nonda*)
- Noni (*Morinda citrifolia*; Rubiaceae)
- Nungu (*Borassus flabellifer*; Arecaceae)
- Nutmeg (*Myristica fragrans*; Myristicaceae)

O

- Oil palm (*Elaeis guineensis*; Arecaceae)
- Olive
- Orange, of which there are sweet (*Citrus sinensis*) and sour (*Citrus aurantium*) species
- Oregon grapc (*Mahonia aquifolium*; Berberidaceae)
- Oroblanco (Citrus paradisi C. grandis) Rutaceae (Also called the sweetie)
- Otaheite gooseberry (*Phyllanthus acidus*; Phyllanthaceae)

P

- Palmyra palm/toddy palm (*Borassus flabellifer*; Arecaceae)
- Papaya (*Carica papaya*; Caricaceae)
- Passion fruit or granadilla (*Passiflora edulis* and other *Passiflora* spp.; Passifloraceae)
- Pawpaw (*Asimina triloba*; Annonaceae, not to be confused with papaya (*Carica papaya*; Caricaceae), which is called pawpaw in some English dialects)
- Peach (of the normal and white variety) and its variant the nectarine (*Prunus persica*)
- Peach palm (*Bactris gasipaes*; Arecaceae)
- Peanut (*Arachis hypogaea*; Fabaceae)
- Peanut butter fruit (*Bunchosia argentea*; Malpighiaceae)
- Pear, European and Asian species (*Pyrus*)
- Pecan (*Carya illinoinensis* or *illinoensis*; Juglandaceae)
- Pentadin (*Pentadiplandra brazzeana Baillon*)
- Pepino (*Solanum muricatum*)
- Pequi (*Caryocar brasiliensis*; Caryocaraceae)
- Persian lime Also known as tahiti lime.
- Persimmon, American, see American persimmon
- Persimmon, oriental, see Oriental persimmon
- Peumo (*Cryptocarya alba*; Lauraceae)
- Phalsa (*Grewia subinaequalis*; Tiliaceae)
- Pigeon pea
- Pigeon plum (*Coccoloba diversifolia*; Polygonaceae)
- Pigface (*Carpobrotus glaucescens*; Aizoaceae)
- Pili nut (*Canarium ovatum*; Burseraceae)
- Pindo palm (*Butia Capitata*; Arecaceae)
- Pineapple (*Ananas comosus* ; Bromeliaceae)
- Pineapple guava (*Feijoa sellowiana*; Myrtaceae)
- Pink-flowered native raspberry (*Rubus parvifolius*; Rosaceae)
- Pistachio (*Pistacia vera*; Anacardiaceae)
- Pitaya, see dragon fruit
- Pitomba (*Eugenia luschnathiana* or *Talisia esculenta*)
- Plum, of which there are several domestic and wild species; dried plums are called **prunes**
- Pluot (Prunus), a Plum/Apricot hybrid, also known as Apriplums, Apriums or Plumcots

- Pois doux (*Inga edulis*, *ice-cream bean*, or *inga-cipó* in Portuguese)
- Pomegranate (*Punica granatum*; Punicaceae)
- Pomelo (also known as the shaddock) (*Citrus maxima*)
- Pommecythère or pomcité (*Spondias cytherea*); also known as golden apple, June plum or Jew plum and ambarella, and as cajamanga in Portuguese
- Pommerac (*Eugenia malaccensis*); also known as Otaheite apple; Malay apple; jambo in Portuguese
- Pond-apple (*Annona glabra*; Annonaceae) Also called alligator-apple and monkey-apple
- Prickly pear (*Opuntia* spp.; Cactaceae) used as both a fruit and vegetable depending on part of plant.
- Pulasan (*Nephelium mutabile*; Sapindaceae)
- Pummelo (*Citrus grandis*; Rutaceae)
- Pumpkin (Cucurbita spp.)
- Pupunha (*Bactris gasipaes*; Arecaceae); also known as *peach plum* or *pewa*
- Purple apple-berry (*Billardiera longiflora*; Pittosporaceae)
- Purple granadilla (*Passiflora edulis f edulis*; Passifloraceae)
- Purple guava (*Psidium rufum*; Myrtaceae)
- Purple mombin (*Spondias purpurea*; Anacardiaceae)

Q

- Quandong (*Santalum acuminatum*; Santalaceae)
- Queensland ebony (*Diospyros humilis*)
- Quince (*Cydonia oblonga* and *Chaenomeles*)

R

- Raisin tree (*Hovenia dulcis*, Rhamnaceae) Also called Japanese raisin tree
- Rambutan (*Nephelium lappaceum*; Sapindaceae)
- Raspberry, several species (genus *Rubus*)
- Red granadilla (*Passiflora coccinea*; Passifloraceae)
- Red mombin (*Spondias purpurea*; Anacardiaceae)
- Red mulberry (*Morus rubra*)
- Rhubarb (*Rheum rhaponticum*; Polygonaceae)
- Riberry (*Syzygium luehmannii*; Myrtaceae), also called lilly pilly, lillipilli, Chinese apple
- Rimu (*Dacrydium cupressinum*)
- Rollinia (*Rollinia mucosa*; Annonaceae)
- Rose apple (*Syzygium jambos*; Myrtaceae)
- Rose hip, the fruitlike base of roses (*Rosa*); used mostly for jams and herbal teas

- Rose-leaf bramble (*Rubus rosifolius*; Rosaceae)
- Rose myrtle (*Archirhodomyrtus beckleri*; Myrtaceae)
- Rough shell macadamia (*Macadamia tetraphylla*; Proteaceae)
- Rowan (*Sorbus*)

S

- Safou (*Dacryodes edulis*), also called *atanga* or *butterfruit*
- Sageretia (*Sageretia theezans*; Rhamnaceae) Also called Mock Buckthorn.
- Saguaro (*Carnegiea gigantea*; Cactaceae)
- Salak (*Salacca edulis*; Arecaceae), also called snakefruit or *cobrafruit*
- Salal (*Gaultheria shallon*; Ericaceae)
- Salmonberry (*Rubus spectabilis*)
- Sandpaper fig (*Ficus coronata*; Moraceae)
- Santol (*Sandoricum koetjape*; Meliaceae)
- Sapote, see mamey
- Sapodilla (*Manilkara zapota*; Sapotaceae), also called chiku, mespel, naseberry, sapadilla, snake fruit, sawo
- Satinash, fibrous, see fibrous satinash
- Saskatoonberry (*Amelanchier alnifolia*, Rosaceae)
- Saw palmetto (*Serenoa repens*; Arecaceae)
- Sea-buckthorn (*Hippophae rhamnoides*; Elaeagnaceae)
- Sea grape (*Coccoloba uvifera*; Polygonaceae)
- Serviceberry or Saskatoon (*Amelanchier*)
- Shipova (× *Sorbopyrus auricularis*)
- Sloe (*Prunus spinosa*, Rosaceae)
- Small-leaf tamarind (*Diploglottis campbellii*; Sapindaceae)
- Snow berry (*Gaultheria hispida*; Ericaceae)
- Soncoya (*Annona diversifolia*; Annonaceae)
- Service tree (*Sorbus domestica*), bears a fruit known as a sorb or sorb apple
- Soursop (*Annona muricata*; Annonaceae), also called guanabana
- Southern crabapple (*Malus angustifolia*; Rosaceae)
- Spanish lime (*Melicoccus bijugatus*; Sapindaceae)
- Spanish tamarind (*Vangueria madagascariensis*)
- Spiny monkey-orange (*Strychnos spinosa*)
- Star apple (*Chrysophyllum cainito*; Chrysobalanaceae), also called caimito or *caimite*
- Starfruit, see carambola

- Strawberry (*Fragaria*)
- Strawberry guava (*Psidium littorale*; Myrtaceae)
- Strawberry myrtle, see ugni
- Strawberry pear
- Sugar apple (*Annona squamosa*; Annonaceae); ata in Portuguese
- Surinam cherry (*Eugenia uniflora*; Myrtaceae) also called Brazilian cherry, Cayenne cherry, pitanga
- Sweet apple-berry (*Billardiera cymosa*; Pittosporaceae)
- Sweet granadilla (*Passiflora ligularis*; Passifloraceae)
- Sweet lemon (*Citrus limetta*)
- Sweet orange
- Sweet pepper
- Sweetsop (*Annona squamosa*; Annonaceae) also called sugar apple
- Sycamore fig (*Ficus sycomorus* Moraceae) also called Old World sycamore] or just sycamore

T

- Tahitian apple (*Spondias dulcis*)
- Tamarillo or tree tomato (*Cyphomandra betacea*; Solanaceae)
- Tamarind (*Tamarindus indica*; Leguminosae)
- Tamarind, Manila, see Manila tamarind
- Tamarind, monkey, see monkey tamarind
- Tamarind, velvet, see velvet tamarind
- Tangerine, and similar
- Tanjong (*Mimusops elengi*; Sapindaceae)
- Texas persimmon (*Diospyros texana*; Ebenaceae)
- Thimbleberry (*Rubus parviflorus*)
- Tōtara (*Podocarpus totara*)
- Toyon (*Heteromeles arbutifolia*; Rosaceae)
- Tropical almond (*Terminalia catappa*; Combretaceae)

U

- Ugni (*Ugni molinae*; Myrtaceae)

V

- Vanilla (*Vanilla planifolia*; Orchidaceae)
- Velvet tamarind (*Dialium indum*; Leguminosae)

W

- Wampee (*Clausena lansium*; Rutaceae)
- Wani (*Mangifera caesia*; Mangifera)
- Water apple (*Syzygium aqueum*; Myrtaceae)
- Watermelon (*Citrullus lanatus*)
- Wax apple ("Syzygium samarangense)
- Wax gourd, or winter melon (*Benincasa hispida*), eaten as a culinary vegetable when mature, but sweet when young
- Wax jambu (*Syzygium samarangense*; Myrtaceae)
- West Indian cherry, see acerola
- White aspen (*Acronychia oblongifolia*; Rutaceae)
- White mulberry (*Morus alba*)
- White sapote (*Casimiroa edulis*; Rutaceae)
- Whortleberry, see bilberry
- Wild grape (*Ampelocissus acetosa*)
- Wild orange (*Capparis mitchellii*; Capparaceae)
- Wild peach (*Terminalia carpentariae*)
- Wild plum (munydjudj) (*Buchanania obovata*)
- Wild plum (*Santalum lanceolatum*)
- Wild mangosteen (*Garcinia indica*)
- Wineberry (*Rubus phoenicolasius*)
- Wolfberry (*Lycium barbarum*, *Lycium* spp.; Solanaceae)
- Wongi (*Manilkara kaukii*; Sapotaceae)
- Wood apple (*Feronia limonia*; Rutaceae)

Y

- Yangmei (*Myrica rubra*; Myricaceae) Also called yumberry, yamamomo, Chinese bayberry, Japanese bayberry, red bayberry, or Chinese strawberry tree
- Yantok, or rattan fruit (*Calamus manillensis*)
- Yellow granadilla (*Passiflora edulis f flavicarpa*; Passifloraceae)
- Yellow mombin (*Spondias mombin*; Anacardiaceae)
- Yellow plum (*Ximenia americana*; Olacaceae)
- Yew (*Taxus baccata*; Taxaceae)
- Youngberry

Z

- Zig zag vine (*Melodurum leichhardtii*; Annonaceae)
- Ziziphus, see jujube

- Zwetschge (*Prunus domestica subsp. domestica*; Rosaceae)

Culinary Fruits by Climate

Temperate fruits

Fruits of temperate climates are almost always borne on trees or woody shrubs or lianas. They will not grow adequately in the tropics, as they need a period of cold (a chilling requirement) each year before they will flower. The apple, pear, cherry, and plum are the most widely grown and eaten, owing to their adaptability. Many other fruits are important regionally but do not figure prominently in commerce. Many sorts of small fruit on this list are gathered from the wild, just as they were in Neolithic times.

Apples

The Pome Fruits

- Apple and crabapple (*Malus*)
- Chokeberry
- Hawthorn

- Loquat
- Medlar
- Pear
- Quince
- Rose hip
- Rowan
- Service tree
- Serviceberry
- Shipova

The stone fruits, drupes of genus *Prunus*

- Apricot
- Cherry, sweet, black, sour, and wild species
- Chokecherry
- Greengage
- Peach varieties and nectarine
- Plum
- Hybrids of the preceding species, such as the plumcot/apriplum, pluot, aprium, and peacotum

Other Temperate Fruits

- Boquila
- Elaeocarpaceae
- Goumi
- Haskap / honeyberry
- Keule
- Lardizabala
- Maqui
- Pawpaw
- Peumo
- Fox grape
- Riverbank grape
- Sand grape
- Amur grape

Berries

In non-technical usage, **berry** means any small fruit that can be eaten whole and lacks objectionable seeds. The **bramble fruits**, compound fruits of genus *Rubus* (blackberries), are some of the most popular of these that are not true berries:

Rubus

Raspberries

- Blackberry, including dewberry, boysenberry, olallieberry, and tayberry
- Cloudberry
- Loganberry
- Raspberry species
- Salmonberry
- Thimbleberry
- Wineberry

True Berries

The true berries are dominated by the family Ericaceae, many of which are hardy in the subarctic:

- Bearberry
- Bilberry
- Blueberry
- Crowberry
- Cranberry
- Falberry

- Huckleberry
- Lingonberry

Other Berries

- Currant
- Elderberry
- Gooseberry
- Hackberry
- Honeysuckle
- Mulberry, including black and yellow mulberry
- Mayapple
- Nannyberry
- Oregon grape
- Sea buckthorn
- Strawberry
- Ugni
- Wolfberry

Mediterranean And Subtropical Fruits

Fruits in this category are not hardy to extreme cold, as the preceding temperate fruits are, yet tolerate some frost and may have a modest chilling requirement. Notable among these are natives of the Mediterranean:

Mediterranean Natives

Grapes

- Olives
- Black mulberry
- Cornelian cherry
- Date
- Fig
- Grapes, Vitis vinifera and raisins
- Jujube
- Pomegranate
- Sycamore fig

Citrus

In the important genus *Citrus* (Rutaceae), some members are tropical, tolerating no frost. All common species of commerce are somewhat hardy:

Lemon

- Blood orange
- Citron
- Clementine
- Grapefruit
- Kumquat
- Lemon
- Lime, including Key, Persian and kaffir lime
- Mandarin
- Naartjie
- Orange (spicy or bitter)

- Pomelo
- Sweet lemon
- Kabosu
- Oroblanco
- Tangerine
- Hybrids of the preceding species, such as the orangelo, tangelo, rangpur, ortanique, lemonade fruit, and ugli fruit

Other Subtropical Fruits

- Avocado
- Carob
- Feijoa
- Guava
- Longan
- Lúcuma
- Lychee
- Passion fruit
- Peanut
- Pond-apple
- Strawberry guava
- Tamarillo
- Yangmei
- Néré

Tropical Fruits

Tropical fruits grow on plants of all habitats. The only characteristic that they share is an intolerance of frost.

A

- Abiu
- Açaí
- Acerola
- Ackee
- African cherry orange
- African moringa
- Agave
- Allspice

- Ambarella
- American persimmon
- Araza
- Atemoya
- Avocado

B

- Babaco
- Bacupari
- Bacuri
- Bael
- Banana
- Barbadine
- Barbados cherry
- Betel nut
- Bignay
- Bilimbi
- Biribi
- Black mulberry
- Black sapote
- Bolivian coconut
- Bottle gourd
- Breadnut
- Burmese grape

C

Caimito fruit cut in half. It is native to the lowlands of Central America and the West Indies

- Caimito
- Calabash tree
- Calamansi
- Camu camu
- Canistel
- Cape gooseberry
- Capulin cherry
- Carambola
- Cassabanana
- Cattley guava
- Cawesh
- Ceriman
- Ceylon gooseberry
- Chayote
- Chempedak
- Chenet
- Cherimoya
- Chilean guava
- Chinese jujube
- Cherry of the Rio Grande
- Chinese olive
- Chupa-chupa
- Coco plum
- Cocona
- Double coconut
- Coconut
- Cola nut
- Costa Rican guava
- Cupuaçu
- Curry-leaf tree
- Cocoplum
- Custard apple

D

- Damson plum
- Date plum
- Dead man's fingers

- Dragonfruit
- Duku
- Durian

E

- Elephant apple
- Emblica

G

- Gambooge
- Genip
- Giant granadilla
- Governor's plum
- Grapes
- Grapefruit
- Grumichama
- Guanabana
- Guarana
- Guava
- Guavaberry

H

- Hairless rambutan
- Hog plum
- Horned melon
- Huito
- honeydew

I

- Ice cream bean
- Ilama
- Imbe
- Indian almond
- Indian fig
- Indian gooseberry
- Indian jujube
- Indian prune

J

- Jaboticaba
- Jackfruit
- Jambul
- Jatobá
- Jelly plum
- Jocote

K

- Kandis
- Kapok
- Karonda
- Kei apple
- Kepel fruit
- Key lime
- Kitembilla
- Kiwano
- Kiwifruit
- Kwai muk
- Korlan
- Kundong

L

- Lakoocha
- Langsat
- Lanzones
- Lemon
- Leucaena
- Limeberry
- Limequat
- Lime
- Longan
- Loquat
- Louvi
- Lucuma
- Lychee

M

- Mabolo
- Macadamia
- Madrono
- Malabar plum
- Malay apple
- Mammee apple
- Mamey
- Mamoncillo
- Mangaba
- Mango
- Mangosteen
- Manila tamarind
- Ma-praang
- Mayan breadnut
- Maypop
- Medlar
- Meiwa kumquat
- Melinjo
- Melon pear
- Miracle fruit
- Monstera
- Montessa granadilla
- Mountain soursop
- Monkey jackfruit
- Monkey tamarind
- Mundu
- Muskmelon

N

- Nagami kumquat
- Nance
- Naranjilla
- Neem
- Noni
- Nutmeg

O

- Oil palm
- Olive
- Otaheite gooseberry
- Orange
- Oriental persimmon

P

- Palmyra palm
- Papaya
- Passion fruit
- Peach palm
- Peanut butter fruit
- Pecan
- Pepino
- Pequi
- Pewa
- Phalsa
- Pigeon pea
- Pili nut
- Pindo palm
- Pineapple
- Pineapple guava
- Pistachio
- Pitaya
- Pitomba
- Pois doux
- Pomegranate
- Pommecythère
- Pommerac
- Pulasan
- Pummelo
- Pupunha
- Purple guava
- Purple granadilla
- Purple mombin

R

- Rambutan
- Red granadilla
- Red mombin
- Riberry
- Ridged gourd
- Rollinia
- Rose apple
- Rough shell macadamia

S

- Safou
- Salak
- Santol
- Sapodilla
- Sea grape
- Soncoya
- Soursop
- Spanish lime
- Star apple
- Strawberry guava
- Strawberry pear
- Sugar apple
- Summer squash
- Surinam cherry
- Sweet granadilla
- Sweet orange
- Sweet pepper
- Sweetsop

T

- Tahitian apple
- Tangerine
- Tamarind
- Tomato

U

- Ugni

V

- Vanilla
- Velvet tamarind
- Voavanga

W

- Wampee
- Water apple
- Watermelon
- Wax jambu
- Wax gourd
- White sapote
- Winged bean
- Wood apple

X

- Xigua

Y

- Yantok
- Yellow granadilla
- Yellow mombin
- Youngberry

Z

- Zucchini

REFERENCES

1. China is black hole of Asia's deforestation, Asia News, 24 March 2008
2. Corlett, R. & Primack, R. (2006). "Tropical Rainforests and the Need for Cross-continental Comparisons". Trends in Ecology & Evolution. 21 (2): 104–110. doi:10.1016/j.tree.2005.12.002.
3. Bruijnzeel, L. A. & Veneklaas, E. J. (1998). "Climatic Conditions and Tropical Montane Forest Productivity: The Fog Has Not Lifted Yet". Ecology. 79 (1): 3. doi:10.1890/0012-9658(1998)079[0003:CCATMF]2.0.CO;2.
4. Phillips, O.; Gentry, A.H.; Reynel, C.; Wilkin, P.; Galvez-Durand b, C. (1994). "Quantitative Ethnobotany and Amazonian Conservation". Conservation Biology. 8 (1): 225–48. doi:10.1046/j.1523-1739.1994.08010225.x.
5. Bourgeron, Patrick S. (1983). "Spatial Aspects of Vegetation Structure". In Frank B. Golley. Tropical Rain Forest Ecosystems. Structure and Function. Ecosystems

of the World (14A ed.). Elsevier Scientific. pp. 29–47. ISBN 0-444-41986-1.

6. Erwin, T.L. (1982). "Tropical forests: Their richness in Coleoptera and other arthropod species" (PDF). The Coleopterists Bulletin. 36: 74–75. JSTOR 4007977.
7. "Sabah". Eastern Native Tree Society. Retrieved 14 November 2007.
8. King, David A. & Clark, Deborah A. (2011). "Allometry of Emergent Tree Species from Saplings to Above-canopy Adults in a Costa Rican Rain Forest". Journal of Tropical Ecology. 27 (6): 573–79. doi:10.1017/S0266467411000319.
9. Denslow, J S (1987). "Tropical Rainforest Gaps and Tree Species Diversity". Annual Review of Ecology and Systematics. 18: 431. doi:10.1146/annurev.es.18.110187.002243.
10. Malhi, Yadvinder & Wright, James (2004). "Spatial patterns and recent trends in the climate of tropical rainforest regions". The Royal Society Biological Sciences. 359 (1443): 311–329. doi:10.1098/rstb.2003.1433.
11. NWS JetStream – Inter-Tropical Convergence Zone. Srh.noaa.gov (5 January 2010). Retrieved on 28 March 2013.
12. Aragao, L. E. O. C. (2009). "Above- and below-ground net primary productivity across ten Amazonian forests on contrasting soils". Biogeosciences. 6 (12): 2759–2778. doi:10.5194/bg-6-2759-2009.
13. Moreira, A.; Fageria, N. K.; Garcia y Garcia, A. (2011). "Soil Fertility, Mineral Nitrogen, and Microbial Biomass in Upland Soils of the Central Amazon under Different Plant Covers". Communications in Soil Science and Plant Analysis. 42 (6): 694–705. doi:10.1080/00103624.2011.550376.
14. Environmental news and information. mongabay.com. Retrieved on 28 March 2013.
15. Cleveland, Cory C. & Townsend, Alan R. (2006). "Nutrient additions to a tropical rain forest drive substantial soil carbon dioxide losses to the atmosphere". PNAS. 103 (27): 10316–10321. Bibcode:2006PNAS..10310316C. doi:10.1073/pnas.0600989103. PMC 1502455 . PMID 16793925.
16. Tang, Yong; Yang, Xiaofei; Cao, Min; Baskin, Carol C.; Baskin, Jerry M. (2010). "Buttress Trees Elevate Soil Heterogeneity and Regulate Seedling Diversity in a Tropical Rainforest" (PDF). Plant and Soil. 338: 301–309. doi:10.1007/s11104-010-0546-4.
17. Sahney, S., Benton, M.J. and Ferry, P.A. (2010). "Links between global taxonomic diversity, ecological diversity and the expansion of vertebrates on land" (PDF). Biology Letters. 6 (4): 544–547. doi:10.1098/rsbl.2009.1024. PMC 2936204 . PMID 20106856.
18. Haffer, J. (1969). "Speciation in Amazonian Forest Birds". Science. 165 (131): 131. Bibcode:1969Sci...165..131H. doi:10.1126/science.165.3889.131.
19. Moritz, C.; Patton, J. L.; Schneider, C. J.; Smith, T. B. (2000). "DIVERSIFICATION OF RAINFOREST FAUNAS: An Integrated Molecular Approach". Annu. Rev. Ecol. Syst. 31: 533. doi:10.1146/annurev.ecolsys.31.1.533.
20. ^ Jump up to:a b c d Barton, Huw; Denham, Tim; Neumann, Katharina; Arroyo-

Kalin, Manuel (2012). "Long-term perspectives on human occupation of tropical rainforests: An introductory overview". Quaternary International. 249: 1–3. doi:10.1016/j.quaint.2011.07.044.

21. ^ Jump up to:a b c d Bailey, R.C., Head, G., Jenike, M., Owen, B., Rechtman, R., Zechenter, E. (1989). "Hunting and gathering in tropical rainforest: is it possible". American Anthropologist. 91 (1): 59–82. doi:10.1525/aa.1989.91.1.02a00040.
22. 'They're killing us': world's most endangered tribe cries for help. The Guardian (22 April 2012). Retrieved on 2016-06-24.
23. Sibaja, Marco (6 June 2012) Brazil's Indigenous Awa Tribe At Risk. Huffington Post
24. González-Ruibal, Alfredo; Hernando, Almudena; Politis, Gustavo (2011). "Ontology of the self and material culture: Arrow-making among the Awá hunter–gatherers (Brazil)". Journal of Anthropological Archaeology. 30: 1. doi:10.1016/j.jaa.2010.10.001.
25. Brazil sees traces of more isolated Amazon tribes. Reuters.com (17 January 2007). Retrieved on 28 March 2013.
26. BBC: First contact with isolated tribes? survivalinternational.org (25 January 2007)
27. Forest peoples in the central African rain forest: focus on the pygmies. fao.org
28. Dufour, D. R. (1990). "Use of tropical rainforest by native Amazonians" (PDF). BioScience. 40 (9): 652–659. doi:10.2307/1311432. JSTOR 1311432.
29. Herrera, Rafael; Jordan, Carl F.; Medina, Ernesto & Klinge, Hans (1981). "How Human Activities Disturb the Nutrient Cycles of a Tropical Rainforest in Amazonia". 10 (2/3, MAB: A Special Issue): 109–114. JSTOR 4312652.
30. Ewel, J J (1986). "Designing Agricultural Ecosystems for the Humid Tropics". Annual Review of Ecology and Systematics. 17: 245. doi:10.1146/annurev.es.17.110186.001333. JSTOR 2096996.
31. Jessup, T. C. & Vayda, A. P. (1988). "Dayaks and forests of interior Borneo" (PDF). Expedition. 30 (1): 5–17.
32. Myers, N. (1985). The primary source, W. W. Norton and Co., New York, pp. 189–193, ISBN 0-393-30262-8
33. Foley, Jonathan A.; Asner, Gregory P.; Costa, Marcos Heil; Coe, Michael T.; Defries, Ruth; Gibbs, Holly K.; Howard, Erica A.; Olson, Sarah; et al. (2007). "Amazonia revealed: forest degradation and loss of ecosystem goods and services in the Amazon Basin". Frontiers in Ecology. 5 (1): 25–32. doi:10.1890/1540-9295(2007)5[25:ARFDAL]2.0.CO;2.
34. Stronza, A. & Gordillo, J. (2008). "Community views of ecotourism: Redefining benefits" (PDF). Annals of Tourism Research. 35 (2): 448. doi:10.1016/j.annals.2008.01.002.
35. Fotiou, S. (October 2001). Environmental Impacts of Tourism. Retrieved 30 November 2007, from Uneptie.org
36. Ismi, A. (1 October 2003), Canadian mining companies set to destroy Ghana's

forest reserves, Canadian Centre for Policy Alternatives Monitor, Ontario, Canada.

37. Walker, Philip L.; Sugiyama, Larry and Chacon, Richard (1998) "Diet, Dental Health, and Cultural Change among Recently Contacted South American Indian Hunter-Horticulturalists", Ch. 17 in Human Dental Development, Morphology, and Pathology. University of Oregon Anthropological Papers, No. 54
38. Tomich, P. T., Noordwijk, V. M., Vosti, A. S., Witcover, J (1998). "Agricultural development with rainforest conservation: methods for seeking best bet alternatives to slash-and-burn, with applications to Brazil and Indonesia". Agricultural Economics. 19: 159–174. doi:10.1016/S0169-5150(98)00032-2.
39. De Jong, Wil; Freitas, Luis; Baluarte, Juan; Van De Kop, Petra; Salazar, Angel; Inga, Erminio; Melendez, Walter; Germaná, Camila (2001). "Secondary forest dynamics in the Amazon floodplain in Peru". Forest Ecology and Management. 150: 135–146. doi:10.1016/S0378-1127(00)00687-3.
40. Semazzi, F. H., Song, Y (2001). "A GCM study of climate change induced by deforestation in Africa". Climate Research. 17: 169–182. doi:10.3354/cr017169.
41. Varghese, Paul (August 2009). "An Overview of REDD, REDD Plus and REDD Readiness" (PDF). Retrieved 23 November 2009.
42. "Hoya Care". Retrieved 2010-06-07.
43. "Cheap banana trees". Archived from the original on June 8, 2010. Retrieved 2010-06-07.
44. "Growing bananas". Retrieved 2010-06-07.
45. "Jungle-Type Bird of Paradise Flowers". Retrieved 2010-06-07.
46. "Fern Care". Retrieved 2010-06-07.
47. "How Not to Kill Your Papaya Tree". Retrieved 2010-06-07.
48. "Orchids – General Care". Retrieved 2010-06-07.
49. Angiosperm Phylogeny Group (2009), "An update of the Angiosperm Phylogeny Group classification for the orders and families of flowering plants: APG III", Botanical Journal of the Linnean Society, 161 (2): 105–121, doi:10.1111/j.1095-8339.2009.00996.x, retrieved 2010-12-10
50. Christenhusz, M. J. M.; Byng, J. W. (2016). "The number of known plants species in the world and its annual increase". Phytotaxa. Magnolia Press. 261 (3): 201–217. doi:10.11646/phytotaxa.261.3.1.
51. Tanaka, N. 2001. Taxonomic revision of the family Cannaceae in the New World and Asia. Makinoa ser. 2, 1:34–43.
52. ^ Jump up to:a b c d e f g h Cooke, Ian. (2001). The Gardener's Guide to Growing Canna, Timber Press. ISBN 978-0-88192-513-5
53. The Cannaceae of the World, H. Maas-van der Kamer & P.J.M. Maas, BLUMEA 53: 247-318
54. Khoshoo, T.N. & Guha, I. - Origin and Evolution of Cultivated Cannas. Vikas Publishing House
55. "canna, n.1." OED Online. Oxford University Press, June 2014. Web. 16 August

2014.

56. Grootjen, C. J. and F. Bouman. 1988. Seed structure in Cannaceae
57. Lerman, J. C. and E. M. Cigliano. 1971. New carbon-14 evidence for six hundred years old Canna compacta seed
58. "Canna indica (canna lilly)", Invasive Plant Compendium, retrieved 5 March 2014
59. FAO, "Canna edulis",http://ecocrop.fao.org/ecocrop/srv/en/cropView?id=2228, accessed 23 Feb 2016
60. de l'Ecluse, Charles (1576) Histoire des plantes rare observées en Espagne
61. Pison (1658), Histoire naturelle du Brésil
62. Lamarck, Jean-Baptiste. Botanical Encyclelopédie.
63. Chaté, E. (1867). Le Canna, son histoire, sa culture. Libraire Centrale d'Agriculture et de Jardinage
64. Aloha 'Aina Jewelry
65. Ganesh Mani Pradhan & Son
66. Plants for a Future
67. Treatment of wastewater by natural systems
68. Constructed wetland for on-site septic treatment.
69. Paul Maas 1985. 195. Cannaceae. & Maas, P. J. M. and H. Maas. 1988. 223. Cannaceae.
70. Species boundaries in Canna (Cannaceae): evidence from nuclear ITS DNA sequence data
71. Percy-Lancaster, S., An Indian Garden. 1927
72. Canna Ehemannii accessdate=25 July 2014
73. Grant, Greg (2011). Heirloom Gardening in the South. College Station, Texas: Texas A&M University Press. p. 187.
74. Canna Synonyms
75. Bailey, L.H. - Canna x generalis. Hortus, 118 (1930); cf. Standley & Steyerm. in Fieldiana, Bot., xxiv. III.204 (1952)
76. Bailey, L.H. - Canna x orchiodes. Gentes Herb. (Ithaca), 1 (3): 120 (1923)
77. Comparison of Crozy & Italian Group cultivars
78. International Code of Nomenclature for Cultivated Plants, C.D. Brickell (Commission Chairman), B.R. Baum, W.L.A. Hetterscheid, A.C. Leslie, J. McNeill, P. Trehane, F. Vrugtman, J.H. Wiersema (eds.). ISBN 90-6605-527-8
79. Hermann, Michael (1995–1996). Crop growth and starch productivity of edible canna. International Potato Center. pp. 295–301.
80. Nobuyuki Tanaka (2004). "The utilization of edible Canna plants in southeastern Asia and southern China". Economic Botany. 58 (1): 112–114. doi:10.1663/0013-0001(2004)058[0112:noep]2.0.co;2. JSTOR 4256780.
81. On the Genus Canna in Yaeyama Islands, the Ryukyus, Japan

82. Edible Canna and its Starch: An Under-Exploited Starch-Producing Plant Resource
83. Progress in the Development of Economic Botany and Knowledge of Food Plants.
84. Hermann, Michael; NK Quynh; D. Peters (1999). "Reappraisal of Edible Canna as a High-Value Starch Crop in Vietnam" (PDF). CIP Program Report 1997-98. Retrieved 27 November 2011.
85. "RHS Plant Selector - Canna 'Alaska'". Retrieved 16 July 2013.
86. "RHS Plant Selector - Canna 'Annaeei'". Retrieved 16 July 2013.
87. "RHS Plant Selector - Canna × ehemanii". Retrieved 16 July 2013.
88. "RHS Plant Selector - Canna 'Erebus'". Retrieved 16 July 2013.
89. "RHS Plant Selector - Canna 'General Eisenhower'". Retrieved 16 July 2013.
90. "RHS Plant Selector - Canna 'Louis Cayeux'". Retrieved 16 July 2013.
91. "RHS Plant Selector - Canna 'Musifolia'". Retrieved 16 July 2013.
92. "RHS Plant Selector - Canna 'Mystique'". Retrieved 16 July 2013.
93. "RHS Plant Selector - Canna 'Phasion'". Retrieved 16 July 2013.
94. "RHS Plant Selector - Canna 'Picasso'". Retrieved 16 July 2013.
95. "RHS Plant Selector - Canna indica 'Russian Red'". Retrieved 16 July 2013.
96. "RHS Plant Selector - Canna 'Shenandoah'". Retrieved 16 July 2013.
97. "RHS Plant Selector - Canna 'Verdi'". Retrieved 16 July 2013.
98. "RHS Plant Selector - Canna 'Withelm Pride'". Retrieved 16 July 2013.
99. "RHS Plant Selector - Canna 'Wyoming'". Retrieved 16 July 2013.
100. Plant Pest handbook
101. Organic Gardening: How to Propagate Canna by Seed.
102. Verhaeghe, P. 2004. Clef des plantes sarmenteuses rustiques en Belgique. Taxonomania 11:1-26
103. "RHS Plant Selector - Hedera algeriensis 'Ravensholst'". Retrieved 20 June 2013.
104. http://asignatura.us.es/abotcam/especies/Hedera_maroccana.html
105. OnlinePlantGuide.com Hedera canariensis /Algerian Ivy. 2007-2015
106. State of California, Department of Public Works. Shrubs most often used. In: California Roadsides, Calif. State Printing Office, 1961.
107. Drill, S.L. et al. S.A.F.E. LANDSCAPES, Southern California Guidebook: Sustainable and Fire-Safe Landscapes In The Wildland Urban Interface. University of California Cooperative Extension, September 2009.
108. Olkowski W. et al. The Potential for Biological Control in Urban Areas. IV Biological Setting for Urban Pest Control. In: Perspectives in Urban Entomology, GW Frankie & CS Koehler, eds. Academic Press, 1978, page 321.
109. "The Plant List: A Working List of All Plant Species". Retrieved 18 December 2014.
110. "USDA GRIN Taxonomy". Retrieved 18 December 2014.
111. Weeds of Australia, Biosecurity Queensland edition: Bear's breeches, Acanthus mollis Archived April 2, 2015, at the Wayback Machine.
112. □κανθος, □κανθα. Liddell, Henry George; Scott, Robert; A Greek–English Lexicon at

the Perseus Project. «άκανθος»., άκανθος (in Greek). Centre for the Greek Language. Harper, Douglas. «acanthus». Online Etymology Dictionary. mollis. Charlton T. Lewis and Charles Short. A Latin Dictionary on Perseus Project.

113. Vitruvius. On Architecture. Book IV.
114. H.A. Wendland in J. Gay, Bulletin de la Société Botanique de France 8:429. 1863(?) («1861»)
115. WCSP, World Checklist of Selected Plant Families: Trachycarpus
116. Flora of China: Trachycarpus fortunei
117. USDA Trachycarpus fortunei

Index